Neurorhetorics

In academia, as well as in popular culture, the prefix "neuro-" now occurs with startling frequency. Scholars publish research in the fields of neuroeconomics, neurophilosophy, neuromarketing, neuropolitics, and neuroeducation. Consumers are targeted with enhanced products and services, such as brain-based training exercises, and babies are kept on a strict regimen of brain music, brain videos, and brain games. The chapters in this book investigate the rhetorical appeal, effects, and implications of this prefix, neuro-, and carefully consider the potential collaborative work between rhetoricians and neuroscientists. Drawing on the increasingly interdisciplinary nature of rhetorical study, *Neurorhetorics* questions how discourses about the brain construct neurological differences, such as mental illness or intelligence measures. Working at the nexus of rhetoric and neuroscience, the authors explore how to operationalize rhetorical inquiry into neuroscience in meaningful ways. They account for the production, dissemination, and appeal of neuroscience research findings, revealing what rhetorics about the brain mean for contemporary public discourse.

This book was originally published as a special issue of *Rhetoric Society Quarterly*.

Jordynn Jack is Associate Professor in the Department of English and Comparative Literature at the University of North Carolina, Chapel Hill, USA. She is author of *Science on the Home Front: American Women Scientists in World War II* (2010), and has written for journals including *Rhetoric Society Quarterly*, *Quarterly Journal of Speech*, and *Rhetoric Review*.

Neurorhetorics

Edited by
Jordynn Jack

Routledge
Taylor & Francis Group

LONDON AND NEW YORK

RHETORIC SOCIETY QUARTERLY

First published 2013 by Routledge

2 Park Square, Milton Park, Abingdon, Oxon OX14 4RN
711 Third Avenue, New York, NY 10017, USA

Routledge is an imprint of the Taylor & Francis Group, an informa business

First issued in paperback 2017

British Library Cataloguing in Publication Data
A catalogue record for this book is available from the British Library

ISBN 13: 978-0-415-52187-1 (hbk)
ISBN 13: 978-1-138-10887-5 (pbk)

Typeset in Minion
by Saxon Graphics Ltd, Derby

Publisher's Note
The Publisher would like to make readers aware that the chapters in this book may be referred to as articles as they are identical to the articles published in the special issue. The Publisher accepts responsibility for any inconsistencies that may have arisen in the course of preparing this volume for print.

Contents

Notes on contributors

L. Gregory Appelbaum is Assistant Professor in the Center for Cognitive Neuroscience at Duke University, USA.

Daniel M. Gross is Associate Professor of English in the School of Humanities at the University of California, Irvine, USA.

David Gruber is Assistant Professor in the English Department at City University of Hong Kong.

Jordynn Jack is Associate Professor in the Department of English and Comparative Literature at the University of North Carolina, Chapel Hill, USA.

John P. Jackson, Jr. is Associate Professor in the Department of Communication at the University of Colorado at Boulder, USA.

Jenell Johnson is Associate Faculty Associate in the Communication Arts Department at the University of Wisconsin at Madison, USA.

Katie Rose Guest Pryal is Clinical Assistant Professor, School of Law, University of North Carolina at Chapel Hill, USA.

Introduction: What are Neurorhetorics?

Jordynn Jack

Imagine sitting down to watch a television documentary about a famous rhetorical persona, say, Martin Luther King or Abraham Lincoln. Instead of the usual commentary from a history professor, though, this special features a new kind of expert: a neurorhetorician named Dr. Aspasia Cranium. Pointing to dramatically colorful brain scans, the neurorhetorician explains how King's "I Have a Dream Speech" activated the brain's "emotion button," located in the nucleus accumbens, leading listeners to connect emotionally to King's message. During the commercial break, you see Dr. Cranium again, this time peddling her line of DVDs and video games, Silver Tongue™ (Unleash Your Rhetorical Power), guaranteed to help you increase your oratorical power through proven neurorhetorical techniques or your money back. By practicing using words to activate the "emotion button" in a simulated brain, you can also learn to activate it in others. If you order in the next thirty minutes, you can get a free, individualized brain map to frame and put up next to your BNr (Bachelor of Neurorhetoric) certificate from Cranium's Neurorhetoric Institute.

Sound farfetched? In the academy, as well as in popular culture, the prefix neuro-now occurs with startling frequency. Scholars now publish research in the fields of neuroeconomics, neurophilosophy, neuromarketing, neuropolitics, and neuroeducation.[1] All of these fields draw from the explanatory power of neuroscience in order to bring new insights to old disciplinary questions. Neuroeconomics, for instance, seeks to use scientific techniques (often neuroscience imaging) to examine how individuals make economic decisions, while neuromarketing seeks to exploit neuroscience insights to trigger the brain's "buy button." Neuropolitics examines the mixing of cultural-political life and the processes of bodies and brains (Connolly xii). Neuroeducation applies insights from neuroscience to generate a better

---

[1] On neuroeconomics, see Glimcher et al., *Neuroeconomics*; Glimcher, *Decisions*; Politser. On neurophilosophy, see Patricia Smith Churchland's *Neurophilosophy* and *Brain-Wise*. On neuropolitics, see Connolly. On neuromarketing, see Renvoise and Morin; Pradeep. On neuroeducation, see Rich and Goldberg. For a rhetorical analysis and critique of how these neuro-disciplines incorporate scientific findings, see Melissa Littlefield and Jenell Johnson's "Lost and Found in Translation" and Papoulias and Callard.

understanding of how students learn. All of these fields have attractive commercial applications. A search of "brain training" books listed on Amazon.com turns up 396 books, with titles such as *Brainfit: 10 Minutes a Day for a Sharper Mind and Memory* (Gediman and Crinella), *Keep Your Brain Alive: 83 Neurobic Exercises* (Katz and Rubin), and *Train Your Brain: 60 Days to a Better Brain* (Kawashima). Websites peddle free brain-based training exercises, and babies are kept on a strict regimen of (doubtfully effective) brain music, brain videos, and brain games. Clearly, the 1980s and 1990s era cultural obsession with physical fitness and body sculpting is being applied, metaphorically, to a new site, the brain, which requires its own set of calisthenics for peak performance. While Michel Foucault defined "technologies of the self" as including operations on "bodies and souls" so as to attain "a certain state of happiness, purity, wisdom, perfection, or immortality" (18), we can now safely add "brains" to the list of targets.

As the parodic opening to this piece suggests, it might be tempting for rhetoric scholars to hop on the neuro-bandwagon. This book might seem to take a similar approach to rhetorical study, an approach that would investigate the "neural correlates" of rhetorical concepts such as pathos, presence, identification, or persuasion. Such an approach might be attractive to rhetoric scholars seeking to draw on the cultural capital of neuroscience, and to those seeking to answer that elusive question about how to study audience response. Collectively, though, the articles in this book argue for an expanded definition of neurorhetorics that acknowledges these impulses, but also upholds the importance of critical and rhetorical perspectives on discourses involving the brain. As we see it, the goal of neurorhetorics—if such a term can be used—would be to investigate the rhetorical appeal, effects, and implications of this prefix, neuro-, as well as to carefully consider collaborative work between rhetoricians and neuroscientists. Drawing on the increasingly interdisciplinary nature of rhetorical study, neurorhetorics would question how discourses about the brain construct neurological difference, determine how to operationalize rhetorical inquiry into neuroscience in meaningful ways, and study what those constructions imply for contemporary public discourse.

In the introduction to their edited book, *Sexualized Brains,* philosophers of science Nicole C. Karafyllis and Gotlind Ulshöfer argue that "neurorhetorics," in practice, often naturalize or construct social classifications, especially along lines of sex and gender (8). Given the human tendency toward both perfection and hierarchy, famously noted in Kenneth Burke's "Definition of Human" (17), these classifications tend to be drawn along traditional lines of social differentiation, demarcating the normal from the abnormal, the heterosexual from the homosexual, the criminal from the law-abiding brain. New classifications have also emerged, such as the depressed brain (Irwin) or the "extreme male" brain (Baron-Cohen) or the "fit," aerobicized brain. Detangling the rhetorical practices that contribute to these brain-based differences might be one task for rhetoric scholars. As rhetorical scholars, we can account for the production, dissemination, and appeal of these social classifications, which draw on scientific as well as popular, cultural, visual, and historical lines of argument.[2] The authors featured in this book offer critical perspectives that may prove particularly useful.

The first part of this book examines how rhetoric scholars might directly engage with the field of neuroscience, which since the 1960s has sought to link psychological studies of the mind with physiological study of the brain. First, the insights of rhetoric of science draw attention to how neurological differences are produced in cognitive neuroscience. In Chapter 1, Jordynn Jack and L. Gregory Appelbaum examine the methodological considerations that shape one particularly prominent scientific approach, functional magnetic resonance imaging, or fMRI. Given the fact that these methodologies are highly contested within cognitive neuroscience, and given the disparities in scientific techniques used to study such concepts as "emotion" or "empathy," Jack and Appelbaum argue that "neurorhetorics" should entail a two-sided approach: the *rhetoric of neuroscience* and the *neuroscience of rhetoric*. In other words, rhetoric scholars should pay careful attention to how cognitive neuroscience is shaped and circulated, rhetorically, in order to make more careful discriminations about what neuroscience might add to our understanding of traditional rhetorical concepts. Jack, a rhetoric scholar and Appelbaum, a neuroscientist, argue that truly interdisciplinary, collaborative research offers one promising approach for studying neurorhetorics responsibly.

In Chapter 2, David Gruber takes up the "neuroscience of rhetoric" angle, examining how such work has proceeded in rhetorical studies. In particular, he examines the rhetorical strategies rhetoric scholars tend to use when they turn to neuroscience to enrich or explain rhetorical theories. Using a case study of one well-received article, Diane Davis' "Identification: Burke and Freud on Who You Are" (published in *Rhetoric Society Quarterly* in 2008), Gruber questions the tendency to employ scientific findings as somehow separable from their own rhetoricity, but notes that this move accomplishes a key rhetorical purpose in the essay: confirming an earlier, Freudian definition of identification and supporting it with newer scientific research.

In Chapter 3, "Toward a Rhetoric of Cognition," Daniel M. Gross offers a response to Jack and Appelbaum's delineation of the area of "Neurorhetorics," arguing that rhetoric scholars should also develop a third area of inquiry, which would focus on how cognitive concepts are experimentally induced, and how rhetoric scholars can offer enriched conceptions of the situationality of those concepts. As a specific instance, Gross employs a widely cited study in which cognitive neuroscientists measured implicit bias against unfamiliar African American faces in White Americans. Gross offers several points of entry for rhetoric scholars who might want to intervene in how this type of study experimentally induces racialized fear, as well as points of affiliation among cognitive scientists

---

[2]Readers should also refer to the position statement published by Gruber et al., which highlights similar concerns, as well as Jan Slaby's essay on "critical neuroscience."

who are themselves questioning how contextual and situational factors shape the concepts they study.

The next three chapters offer three methods for rhetorical examination of those contextual and situational factors that are evoked in neuroscience studies. An historical perspective shows that, while the prefix neuro- has recently proliferated in usage, there's nothing particularly new about arguments that construct differences based on brain biology. In Chapter 4, John P. Jackson shows that controversies over how human brains can be classified persisted over the course of the twentieth-century, often by drawing on physical measurements such as the "cephalic index" in order to rank humans by race. Jackson argues that rhetorical concepts drawn from jurisprudence, such as "burden of proof," can illuminate how such arguments are carried out. Anthropologist Franz Boas, arguing against the racial hierarchies constructed via such brain measurements successfully shifted the burden of proof onto those who wished to uphold those hierarchies. Yet, Jackson shows, those wishing to revive racist theories of intelligence and brain capacity often argue by attempting to shift the burden of proof back onto those who would disprove their theories, demonstrating the volatility and fluidity of who bears the burden in scientific controversies about just what we can surmise from brain-based studies. As Gross mentions in his chapter, a rhetorical perspective encourages researchers to consider how racialized differences evoked in scientific studies have been historically produced and circulated; Jackson offers one approach to investigating those differences.

The next two articles in this issue offer methods for studying a different type of neurological difference, mental illness. In Chapter 5, Jenell Johnson draws on insights of disability studies to examine how stigmatization functions rhetorically to demarcate individuals with mental illness. Her case study examines Thomas Eagleton's brief stint as Democratic vice Presidential Nominee, which ended when his history of depression, shock therapy, and hospitalization was made public. Notably, discourse about Eagleton's illness focused not only on his history, but on his public performance, with commentators scrutinizing Eagleton's physical appearance and gestures for signs of depression. Johnson concludes that the process of stigmatization is a rhetorical one, one that has important consequences for agency, representation, and power. A deeper understanding of depression in neuroscience studies would involve attention to these rhetorical and situational constructions of depression, rather than positioning it as an ahistorical, static condition.

Katie Rose Guest Pryal's article draws on narrative and genre theory to show how neurorhetorics tend to construct disabilities, and individuals' experiences thereof, along certain storylines and tropes, in this case in ways that may shape experience of disability as well as available lines of authority. From such a perspective, the genres of neurorhetoric become an important site of inquiry. Pryal presents a case study of the genre she calls "mood memoirs," which offer rhetorical space for individuals with mood disorders to claim authority, argue against medicalized interpretations, and construct new narratives of mental disability. Narrative theory offers an additional avenue to understand, especially, how people with neurological differences speak back to the dominant, medicalized and scientific narratives that often limit their rhetorical authority. As is the case with depression, mood disorders might be more fully theorized in neuroscience studies if researchers attend to the contexts in which they are invoked, including the genre of the experiment or experimental report as well as memoir and autobiography.

Of course, a number of additional approaches to neurorhetorics might be mentioned. Not included here, but important for this area of researchers, would be studies of visual rhetorics, of popularization, and approaches drawing on feminist theory, critical theory, and other interdisciplinary investigations. Briefly, visual rhetorics might help to account for the persuasive appeal of neuroscience images, as well as for visual representations of disability (upon which Jack and Appelbaum and Johnson both touch). Studies of popularization would help to account for the diffusion of neurorhetorics in popular magazines and newspapers, and in commercial applications. Feminist and gender studies might account for the continuing trend of constructing neurological difference along the lines of sex/gender (see Condit). For instance, Jack has examined the appeal of the "extreme male brain," posited as a new social classification explaining autism and Asperger's disorder (Jack, "The Extreme Male Brain?"). The production or exclusion of lesbian, gay, bisexual, and transgendered brains should also be of interest to neurorhetoricians. Critical theories might examine the rhetorical production of "technologies of the self" (Foucault 18) in, for example, brain-training computer games, self-help books, and the like; as Davi Johnson has demonstrated in her analysis of brain-based self-help books, popular discourses based on neuroscience "encourage particular kinds of selves, or particular types of citizens, who are more or less amenable to diverse political agendas" (148). Keeping in mind Foucault's assertion that technologies of the self rarely function without technologies of production and power (18), rhetoric scholars might question how constructions of neurorhetorics serve the interests of late-capitalism, which values feminized emotional skills (as in Daniel Goleman's notion of Emotional Intelligence) if not feminized bodies, even as it valorizes the "extreme male brain" as a metonym for the knowledge economy. These discourses may result not in a greater proportion of women in the knowledge economy, but new constructions of masculinity that stress geek prowess alongside communication and teamwork skills.[3] We might pay attention to how arguments from neuroscience legitimate juridical and political power, in technologies such as lie detection (Littlefield), especially in a post-9/11 context. (Perhaps the obsession with the criminal brain of the early twentieth century continues, now in the guise of the "terrorist brain"?) The growth of animal studies might prompt researchers to question how the animal brain is argued into

---

[3]See Cooper.

place as, on the one hand, a model for human brains and, on the other hand, as a limit case against which uniquely human capabilities can be posited. In short, neurorhetorics should prove to be a diverse, interdisciplinary field, one that offers up a range of critical questions and approaches.

Fundamentally, though, what unifies this field should be a prudential approach in our understanding and use of the term neurorhetoric. As Jack and Appelbaum point out, such studies can easily fall into traps of neuro-realism, neuro-essentialism, and neuro-policy, all of which can tend towards uncritical fetishization of the brain as a scientific object divorced from its historical and rhetorical context. The work of neurorhetorics should, therefore, be a cautious and disciplined one, working through the tangle of discourses, claims, and arguments that often seek to reinstate or exaggerate brain differences, limit some individuals from rhetorical participation, or offer new technologies for the human tendency towards perfection and domination.

References

Baron-Cohen, Simon. "The Extreme-Male-Brain Theory of Autism." *Trends in Cognitive Science* 6.6 (2002): 248-54. Print.

Burke, Kenneth. *Language as Symbolic Action*. Berkeley: University of California Press, 1966. Print.

Condit, Celeste. "How Bad Science Stays That Way: Of Brain Sex, Demarcation, and the Status of Truth in the Rhetoric of Science." *Rhetoric Society Quarterly* 26 (1996): 83-109. Print.

Cooper, Marianne. "Being the 'Go-to Guy': Fatherhood, Masculinity, and the Organization of Work in Silicon Valley." *Qualitative Sociology* 23.4 (2000): 379-405. Print.

Foucault, Michel. "Technologies of the Self." *Technologies of the Self: A Seminar with Michel Foucault*. Eds. Luther H. Martin, Huck Gutman, Patrick H. Hutton. Amherst: University of Massachusetts Press, 1988. 16-49. Print.

Gediman, Corinne L., and Francis M. Crinella. *Brainfit: 10 Minutes a Day for a Sharper Mind and Memory*. Nashville: Rutledge Hill Press, 2005. Print.

Glimcher, Paul W. *Decisions, Uncertainty, and the Brain: The Science of Neuroeconomics* Cambridge, MA: MIT Press, 2008. Print.

Glimcher, Paul W., et al., eds. *Neuroeconomics: Decision Making and the Brain*. London; San Diego: Academic Press, 2008. Print.

Goleman, Daniel. *Emotional Intelligence*. New York: Bantam Books, 1996. Print.

Gruber, David et al. "Rhetoric and the Neurosciences: Engagement and Exploration." *POROI: Project on Rhetoric of Inquiry* 7.1 (2011). Web. 10 July 2012. http://ir.uiowo.edu/poroi/vol7/iss1/11.

Irwin, W. et al. "Amygdalar Interhemispheric Functional Connectivity Differs between the Non-Depressed and Depressed Human Brain." *NeuroImage* 21.2 (2004): 674-86. Print.

Jack, Jordynn. "The Extreme Male Brain?' Incrementum and the Rhetorical Gendering of Autism." *Disability Studies Quarterly* 31.3 (2011), web, special issue on Rhetoric and Disability.

Johnson, Davi. "'How Do You Know Unless You Look?': Brain Imaging, Biopower and Practical Neuroscience." *Journal of Medical Humanities* 29.3 (2008): 147-61. Print.

Johnson, Jenell and Melissa Littlefield. "Lost and Found in Translation: Popular Neuroscience in the Emerging Neurodisciplines." *Advances in Medical Sociology* 13 (2011): 279-297.

Karafyllis, Nicole C., and Gotlind Ulshöfer. "Introduction: Intelligent Emotions and Sexualized Brains--Discourses, Scientific Models, and Their Interdependencies." *Sexualized Brains, Scientific Modelling of Emotional Intelligence from a Cultural Perspective*. Eds. Karafyllis, Nicole C. and Gotlind Ulshöfer. Cambridge, MA: MIT Press, 2008. 1-49. Print.

Katz, Lawrence, and Manning Rubin. *Keep Your Brain Alive: 83 Neurobic Exercises*. New York: Workman Publishing Company, 1998. Print.

Kawashima, Ryuta. *Train Your Brain: 60 Days to a Better Brain*. Teaneck, NJ: Kumon Publishing North America, 2005. Print.

Littlefield, Melissa. "Constructing the Organ of Deceipt: The Rhetoric of fMRI and Brain Fingerprinting in Post-9/11 America." *Science, Technology, & Human Values* 34 (2009): 365-92. Print.

Papoulias, C., and F. Callard. "Biology's Gift: Interrogating the Turn to Affect." *Body & Society* 16.1 (2010): 29-56. Print.

Politser, Peter. *Neuroeconomics: A Guide to the New Science of Making Choices*. Oxford: Oxford University Press, 2008. Print.

Pradeep, A. K. *The Buying Brain: Secrets for Selling to the Subconscious Mind*. New York: Wiley 2010. Print.

Rich, Barbara, and Johanna Goldberg, eds. *Neuroeducation: Learning, Arts, and the Brain*. New York; Washington, DC: Dana Press, 2009. Print.

Slaby, Jan. "Steps Towards a Critical Neuroscience." *Phenomenology and Cognitive Sciences* 9 (2010): 397-416. Print.

"This is Your Brain on Rhetoric": Research Directions for Neurorhetorics

Jordynn Jack & L. Gregory Appelbaum

Neuroscience research findings yield fascinating new insights into human cognition and communication. Rhetoricians may be attracted to neuroscience research that uses imaging tools (such as fMRI) to draw inferences about rhetorical concepts, such as emotion, reason, or empathy. Yet this interdisciplinary effort poses challenges to rhetorical scholars. Accordingly, research in neurorhetorics should be two-sided: not only should researchers question the neuroscience of rhetoric (the brain functions related to persuasion and argument), but they should also inquire into the rhetoric of neuroscience (how neuroscience research findings are framed rhetorically). This two-sided approach can help rhetoric scholars to use neuroscience insights in a responsible manner, minimizing analytical pitfalls. These two approaches can be combined to examine neuroscience discussions about methodology, research, and emotion, and studies of autism and empathy, with a rhetorical as well as scientific lens. Such an approach yields productive insights into rhetoric while minimizing potential pitfalls of interdisciplinary work.

At a time when cultural critics lament declining popular interest in science, neuroscience research findings are only gaining in popularity. Highly persuasive neuroscience-related findings are touted for their potential to transform advertising, political campaigns, and law (for example, through new brain-based "lie detectors").[1] Those hoping to improve their own brains can read self-help books, play "brain training" computer and video games, listen to specially designed meditations, and train their children's brains with Baby Einstein, Beethoven for Babies, and similar devices.[2]

---

[1] For a rhetorical-cultural analysis of brain-based lie detectors, see Littlefield.

[2] The scientific evidence for these devices varies considerably. For instance, one 2006 study suggested that each hour of television or video viewing (regardless of type) was actually associated with a 16.99-point *decrease* in MacArthur-Bates Communicative Development Inventory CDI score, an indicator of early language proficiency. See Zimmerman et al.

Neuroscientific research findings are reported in mainstream news outlets with striking regularity. Through scientific and technical developments, researchers can now track active neural systems and document the relationship between brain chemistry, human behavior, and mental activities. These undertakings seem to offer concrete, material proof of concepts previously considered ephemeral, especially when claims are supported with showy, multicolored brain scan images.[3]

In rhetorical studies, there seem to be two main approaches to studying this bourgeoning attention to all things *neuro-*. One area of study under the rubric of neurorhetorics might be *the rhetoric of neuroscience*—inquiry into the modes, effects, and implications of scientific discourses about the brain. To take up a recent example, on 3 February 2010, a Reuters news report featured the following headline: "Vegetative patient 'talks' using brain waves" (Kelland). According to reports carried in nearly every major news outlet, British and Belgian researchers used functional magnetic resonance imaging (fMRI) to demonstrate that a comatose man was able to think "yes" or "no," intentionally altering his brain activity to communicate with the researchers. Newspapers and magazines reprinted the dramatic images of brain activation that appeared in the original scientific report in the *New England Journal of Medicine*, with "yes" answers featuring orange and "no" answers showing blue spots. The findings immediately prompted debates in popular venues. As is often the case with widely reported neuroscience findings, this announcement reinvigorated public arguments about medical care, govern-mentality, and the politics of life itself. Rhetoric scholars should certainly pay attention to how scientific appeals function in these debates.

A second approach might be *the neuroscience of rhetoric*, drawing new insights into language, persuasion, and communication from neuroscience research. Find-ings such as this study of noncommunicative patients can prompt us to broaden our very definitions of rhetoric to include those with impaired communication (such as autism, aphasia, or "locked-in syndrome"), asking how communication occurs through different means, or how brain differences might influence communication. Cynthia Lewiecki-Wilson argues that "we need an expanded understanding of rhetoricity as a potential, and a broadened concept of rhetoric to include collaborative and mediated rhetorics that work with the performative rhetoric of bodies that 'speak' with/out language" (157). Surely, cognitive neu-roscience findings can play an important role in such an endeavor. Neuroscience findings might also add new insights to longstanding rhetorical issues, such as the relationship between *pathos* and *logos*, or emotion and logic, or other cognitive dimensions of rhetoric (Flower; Arthos; Oakley). Indeed, Mark Turner goes so far as to suggest that "If Aristotle were alive today he would be studying this [neuroscience] research and revising his work accordingly" (10).

---

[3]See, for instance, Mooney and Kirshenbaum; Specter.

In this article, we, a neuroscientist and a rhetoric-of-science scholar, argue that the rhetoric of neuroscience and the neuroscience of rhetoric should be intertwined. In other words, to work with neuroscience research findings one should carefully analyze that work with a rhetorical as well as a scientific lens, paying attention to the rhetorical workings of accounts of cognitive neuroscience research. Rhetoricians who would like to do work in neurorhetorics should understand how knowledge is established rhetorically and empirically in the field of cognitive neuroscience, how to interpret scientific findings critically, and how to avoid pitfalls of interpretation that could lead to misleading arguments about rhetoric. Here we demonstrate the kinds of considerations rhetoric scholars should use to examine neuroscience research. First, in order to highlight the complex methodological choices that go into neuroscience research studies, we introduce a contentious debate concerning common analytical practices for functional magnetic resonance imaging. To give rhetoric scholars a set of tools for understanding these complex arguments, we highlight key topoi scientists use to negotiate methodological argument, such as accuracy, efficiency, and bias. Second, we examine how neuroscience researchers define key concepts that may also be of interest to rhetorical scholars, such as emotion, reason, and empathy, considering whether those definitions square with traditional rhetorical concepts of pathos, logos, and identification. In the third section, we consider how a single research article in neuroscience is framed rhetorically, including how decisions about terminology, research questions, and research subjects are rhetorical as well as empirical decisions. In the final section we identify common tropes used in popular accounts of neuroscience research findings. We offer guidelines in each section for rhetorical scholars who would like to work with neuroscience findings, and conclude by offering a set of suggested topics for future research that can constitute what we call neurorhetorics.

Accuracy, Bias, and Efficiency: Methodological Topoi in Human Brain Imaging

As scholars in the rhetoric of science have demonstrated, research findings are shaped rhetorically to fit with scientists' shared expectations. As Lawrence Prelli has argued, scientists use "an identifiable, finite set of value-laden topics as they produce and evaluate claims and counterclaims involving community problems and concerns" (5). Some of these topics (or *topoi*) include accuracy (200), quantitative precision (195), and bias. The accuracy *topos* focuses on the degree to which methods, procedures, and statistical calculations match what is being measured, while the precision *topos* focuses attention on the degree of reliability of the experimental method. Bias refers to the potential for the results to be influenced by factors unrelated to the variable being tested.

In the case of neuroscience, researchers use these three topoi to argue for methods that can usefully extend existing knowledge of the brain's structures and functions. One approach involves using case studies of individuals with brain deficits to draw inferences about normal brain functions. A second approach requires careful, statistical analysis of digitized data generated through imaging technologies such as fMRI or positron emission tomography (PET) (Beaulieu "From Brainbank"). As Michael E. Lynch explains, this data becomes visible through various technologies that transform specimens (animal or human brains) such that "[t]he squishy stuff of the brain becomes a subject of graphic comparison, sequential analysis, numerical measure, and statistical summary" (273). The methods used to accurately extract data from squishy brains are rhetorically negotiated through ongoing debates.

In order to understand these debates, a brief overview of neuroimaging research techniques is important. Through recent advancements in fMRI capabilities, researchers have been able to gain advanced understanding of the activity, structure, and function of the human brain on a fine spatial scale (Bandettini; Poldrack et al.). In most instances, the primary objective in acquiring fMRI data is to infer information about the brain activity that supports cognitive functions (such as perception, memory, emotion) from local changes in blood oxygen content. Increases in neural activity cause variations in blood oxygenation, which in turn cause changes in magnetization that can be detected in an MRI scanner. While these changes (called Blood Oxygenation Level Dependent or BOLD activity) offer a somewhat indirect measure of neural activity, they are widely accepted as a close proxy for the synaptic activity assumed to underlie neuronal communication, brain function, and ultimately cognition (Logothetis and Wandell; Logothetis et al.; Bandettini).[4]

In the hands of cognitive neuroscientists, an fMRI experiment is typically carried out by presenting a subject with a stimulus (such as an image, word problem, or even scent) and a task that requires some kind of response (answering a simple multiple choice question, choosing yes or no, etc.). Neuroscientists analyze the resulting data with regard to specific experimental contrasts designed to isolate, in a meaningful way, specified cognitive functions (e.g., subtraction between remembered and forgotten items from a list). As a result of a single experimental session, researchers can identify minute, specific regions of BOLD activation that correlate with the task at hand in one individual's brain.[5] However, given the inherent variability between individuals in brain anatomy, these activations can not easily be generalized across individuals. The activation patterns may not land consistently in the same place in different brains, nor can they be defined by any

---

[4]Scholars hoping to work with fMRI research findings might wish to consult a textbook explaining basic methodological procedures, such as Scott A. Huettel's *Functional Magnetic Resonance Imaging*.

[5]While this is typical, not all fMRI experimental designs test hypotheses about the specialization of localized regions of the brain. For example, a large number of recent papers have focused on decoding the information that is represented across the whole brain at a particular point in time to a particular class of stimuli.

All science is rhetoric

set of standard anatomical co-ordinates (see Saxe et al. 2006). In order to draw conclusions about brains in general, and not about single individuals, neuroscientists need to establish some basis of comparison across brains, even though they differ in anatomy, size, and arrangement. This is where methodological arguments come in, since neuroscientists must argue for the accuracy and efficiency of their preferred techniques for addressing this challenge.

One approach involves acquiring information from separate "localizer" scans in each subject. Neuroscience researchers Rebecca Saxe, Matthew Brett, and Nancy Kanwisher argue such an approach can "constrain the identification of what is the same brain region across individuals," allowing researchers to more easily "combine data across subjects, studies, and labs" (1089). In rhetorical terms, these researchers argue from the accuracy topos. By identifying regions that function similarly across subjects, they claim that localizer scans allow for more accurate representations of how the brain works. In addition, Saxe, Brett, and Kanwisher argue from efficiency and bias, claiming that the functional regions-of-interest (fROI) approach allows researchers to "specify in advance the region(s) in which a hypothesis will be tested," which "increases statistical power by reducing the search space from tens of thousands of voxels to just a handful of ROIs" (1090). In contrast, the authors claim that whole-head comparisons will "produce an explosion of multiple comparisons, requiring powerful corrections to control false positives" (1090). In this way, they position the fROI approach as more accurate, more efficient, and less likely to lead to biased results (such as false positives). By bias, they mean statistical bias (not personal bias), which can result simply from taking multiple measurements of the whole head. Given the complexity of the brain and the sheer number of neurons it contains, some voxels might indicate brain activity that appears to correlate with the task in question, but that is actually due to sheer chance. In debates about fMRI methodology, the accusation that one technique or another might lead to more false positives serves as a way to position that technique as less sound than the preferred technique.

Using functional localizers, or fROIs, represents a dramatic shift away from more traditional analytical approaches that take into account all measurements from the whole recorded volume, so-called whole-head measurements. Notably, those who support a whole-head approach argue from the very same topoi as those who argue for the fROI approach. For instance, Karl Friston is a vocal proponent of the whole-head approach, which he claims allows for greater accuracy precisely because it does not pinpoint a region of interest *a priori* (Friston et al.; Friston and Henson). Friston points out that the only way to guarantee one has not overlooked potentially interesting activations is to test every voxel (the 3-D unit of measurement in fMRI), a tactic that cannot be done by limiting analysis to only those areas pre-defined in a localizer scan. Drawing on the efficiency topos, Friston et al. argue that whole-brain approaches provide "increased statistical efficiency," making it possible to report results for all locations in the brain while statistically accounting for the multiple tests performed across the whole volume

(Friston et al. 1086). In their defense of the whole-head approach, Friston et al. also argue from the topos of bias, claiming that in the whole-head approach, "the test for one main effect cannot bias the test for other main effects or interactions" (Friston and Henson 1098).

As is the case with any scientific method, claims based on fMRI data rely on chain of inferences that link the data to the psychological function or construct of interest. Each step of this chain raises potential questions about the inferences that can be garnered from the data. The nature and meaning of data are in turn shaped by a series of methodological and conceptual choices made by scientists. This ongoing debate regarding the appropriate tactics to use in fMRI data analysis highlights the fact that neuroscientists have not yet established consensus on these underlying assumptions. It is therefore up to the author to adequately communicate their methodology (Poldrack et al.) and to the reader to be versed in the meaning, trends, and nuances of the methodologies employed.

For researchers hoping to discover new insights into rhetoric and communication from brain studies, it might be tempting to lump together a number of research findings on a topic (such as desire or reason). Yet, each of those studies, individually, might use a different technology (such as PET vs. fMRI), employ a different methodology (such as fROI or whole-brain analysis), and use different kinds of stimuli to evoke a given mental state (images, sounds, smells, etc.). To draw conclusions from such a disparate group of studies requires significant technical knowledge. While rhetoric scholars might find neuroscience methods difficult to understand, they can start by paying attention to these topoi. By looking for terms such as "false positive," "bias," or "assumptions," rhetoric scholars can ferret out places where neuroscientists argue for their methods (or argue against others).

Same Words, Different Meanings: Neurorhetorics of Reason and Desire

Rhetorical scholars have long held a principal interest in reason, emotion, and how they work together to achieve persuasion. These fundamental aspects of human behavior have recently emerged into a rapidly growing branch of empirical neuroscience, called neuroeconomics. As the name implies, neuroeconomics employs both neuroscience techniques and economic theory to test how desire, reason, and choice are represented in the human mind, and, ultimately, why humans make the choices that they do. Neuroeconomics may therefore hold a particularly promising avenue for rhetorical scholars to explore questions that have traditionally been tied to verbal appeals: how people are ultimately persuaded toward a particular course of action.

Rhetoric scholars might be particularly interested in how terms like emotion and reason (which evoke the ancient rhetorical proofs, *pathos* and *logos*) can be studied experimentally in neuroeconomics. In this way, we might gain a deeper understanding of what parts of the brain are activated by emotional stimuli (such as memories of events that signal threat) or by reasoning tasks (such as decision/reward tasks involving the anticipation of gains and losses) (Labar; Carter et al.). Nevertheless, neuroeconomics must be approached with care, since reason and emotion can be difficult concepts to pin down. In this section, we examine how researchers in neuroeconomics understand reason and emotion, how they operationalize those qualities in experiments, and how those understandings do or do not line up with how rhetoricians understand reason and emotion.

Of course, the word "neuroeconomics" itself suggests that the field draws on a specific understanding of human action, one that frames such issues primarily in economic terms. The assumption underlying much of this research is that humans make decisions according to calculations of rewards, risk, and value, and that these are represented in concrete and testable psychological and neural terms. If the brain is responsible for carrying out all of the decisions that humans make, understanding the physiological functions of the brain will help explain why people make specific choices and why they often fail to make optimal decisions.

The interplay between such theory and neurobiology has led to productive insights. Over the past several years, neuroscientists have begun to identify basic computational and physiological functions that explain how reasoning works. One common model is a compensatory one, where individuals make decisions based on calculations of positive versus negative outcomes (Rangel). In this model, decision makers must first form mental representations of the available options, and then assign each option some value according to a common currency (such as monetary gain). Next, the organism compares the values of different options and chooses a specific course of action. After the action is completed, the organism measures the benefit gained, and this information is fed back into the decision mechanism to improve future choices.

A growing body of neuroscientific evidence supports this framework. For example, researchers have found that some neurons in the brain adjust their firing rate with the magnitude and probability of reward (Platt and Glimcher). Similarly, researchers have shown that neurons in the monkey orbitofrontal cortex encode the value of goods (Padoa-Schioppa and Assad), while others have suggested that the frontal cortex neurons represent decision variables such as probability, magnitude, and cost (Kennerley et al.). Collectively, this evidence suggests that subjective value is represented in the nervous system, and that individuals make choices by weighing these values. In this model, decisions are made primarily through rational calculations of value, with the goal being for organisms to maximize their reward (whether it be money, food, or something else).

But does this understanding of reason and emotion line up with the assumptions rhetorical scholars might make about those concepts, which since Aristotle's *Rhetoric*, have been associated with logos and pathos? We might be tempted to take these studies as outside proof that such concepts exist, or to suggest the possibility of someday teasing out rhetorical appeals scientifically (an idea being implemented in the field of neuromarketing). Yet, rhetorical scholars should be careful to distinguish our own understandings of emotion and logic from those supposed by neuroscientists. Daniel Gross argues that Aristotle understands the passions as a sort a "political economy," but the emotions in this theory are decidedly public and rhetorical (6). Anger, for Aristotle, "is a deeply social passion provoked by perceived, unjustified slights," presupposing "a public stage where social status is always insecure" (2). According to Gross, emotions that were at one time treated as "externalized forms of currency" have been folded into the brain, where they are now understood as hardwired and biological, not political and rhetorical (8). While the notion of an emotional economy might appear in both fields, then, the nature of that economy varies significantly.

To determine how, exactly, neuroscientific understandings of reason and logic might match up with rhetorical ones, we searched PubMed for articles that contained the terms *reason, emotion,* and *fMRI.* Out of 83 articles, we chose 20 that attempted to track individuals' response to emotional stimuli and/or reasoned judgments. For each article (see Appendix), we tracked whether or not definitions were provided for the terms *reason* and *emotion,* noted what definitions were given, and determined how those fuzzy concepts were *operationalized,* or rendered scientifically measurable. The studies we selected focused on such topics as gender differences in cognitive control of emotion, the "neural correlates" of empathy, and the recruitment of specific brain regions in inductive reasoning. Obviously, direct comparison of these articles is impossible, and that is not our intent. Our aim was not to conduct an exhaustive study of how scientists operationalize these concepts, but simply to get a preliminary sense of how scientific understandings of reason and emotion might square with rhetorical conceptions.

In many cases, researchers did not define what was meant by key terms such as *emotion* or *reason*—only six of the articles in our sample did so. Perhaps the writers assumed that their readers already shared a common, disciplinary definition. For non-neuroscientists, then, this poses a challenge: what do the authors mean by a term like *emotion* if it is not defined? Is there a standard definition or understanding about this term as it is used in the field? And might these definitions differ between sub-fields?

When definitions were given, they varied in format and content. For instance, studies of reason usually offered provisional definitions, as in these four:

- reasoning "combines prior information with new beliefs or conclusions and usually comes in the form of cognitive manipulations...that require working memory" (Schaich Borg et al. 803)
- "a combination of cognitive processes that allows us to draw inferences from a given set of information and reach conclusions that are not explicitly available, providing new knowledge" (Canessa et al. 930)

- "By 'reasoning,' we refer to relatively slow and deliberative processes involving abstraction and at least some introspectively accessible components" (Greene et al. 389)
- "Inductive reasoning is defined as the process of inferring a general rule (conclusion) by observation and analysis of specific instances (premises). Inductive reasoning is used when generating hypotheses, formulating theories and discovering relationships, and is essential for scientific discovery." (Lu et al. 74)

Anyone hoping to draw conclusions about reason as an element of rhetoric would need to take into account these differing definitions, weighing whether or not they are similar enough to warrant generalizations to rhetorical study. While rhetoricians may wish to associate reason with *logos*, none of these articles considers how individuals are persuaded by logical arguments. In these studies, participants are usually presented with logical puzzles or problems they must solve individually. It would be difficult for a rhetorical scholar to draw clear inferences about logical persuasion from these studies, since they do not focus specifically on how the brain responds to logical appeals.

In the studies mentioning empathy, one cited *Encyclopedia Britannica*'s definition of empathy—"the ability to imagine oneself in another's place and understand the other's feelings, desires, ideas, and actions"—along with criteria from a previous study (Krämer et al. 110). A second defined empathy as "the capacity to share and appreciate others' emotional and affective states in relation to oneself," drawing on previous work by other researchers (Akitsuki and Decety 722). While these definitions are similar, in the first one, empathy involves propelling oneself outward into another's "place," while the second involves the opposite movement of considering another's emotions "in relation to oneself"—the first is outer-directed, the second inner-directed.

For rhetoric scholars, the next step might be to consider how these definitions compare to rhetorical ones. Both of the definitions cited here envisioned empathy as an ability or capacity, something one presumably either has or does not have. On the face of it, these definitions might square with Quintilian's notion that the most effective rhetors possess a capacity to feel the emotions they seek to evoke (Quintilian 6.2.26). For Quintilian, though, empathy is a distinctly performative skill, since orators who can "best conceive such images will have the greatest power in moving the feelings" (6.2.29). In his formulation, empathy represents a capacity to conjure for oneself the emotional states that move the feelings, and to project those emotional states to an audience. Alternately, we might be tempted to line up these fMRI studies with Kenneth Burke's concept of identification, which suggests that "You persuade a man only insofar as you can talk his language by speech, gesture, tonality, order, image, attitude, idea, *identifying* your ways with his" (Burke 55, his emphasis). In any case, both Quintilian and Burke add a dimension to empathy that is lacking in the scientific accounts— the capacity not only to put oneself in another's shoes, but then to *take on* or perform that person's emotions, to "talk his language," as Burke suggests, or to paint an image that evokes those emotions, in Quintilian's conception.

In order for an fMRI study of empathy to map neatly onto these definitions, the study would have to operationalize this specific, rhetorical definition of empathy—not just any study of empathy will necessarily apply. The choice of stimulus would also be significant. In our sample, emotion was evoked using images of neutral or emotional faces (Kompus et al.), faceless cartoons in emotional or neutral situations (Krämer et al.), negative olfactory stimulation (by means of rotten yeast) (Koch et al.), and angry or neutral voices reading nonsense utterances (Sander et al.). Only in a few cases did studies focusing on emotion involve subjects reading or listening to meaningful text—usually a few lines only (Harris, Sheth, and Cohen; Ferstl and von Cramon; Schaich Borg et al.). To date, no fMRI studies that we could find studied individuals' neuronal responses to explicitly rhetorical stimuli—there have been no "this is your brain on Martin Luther King's 'I Have a Dream' speech" studies (although perhaps it is only a matter of time before such a study appears).

In the remaining articles, the terms *emotion* or *reason* were either taken as given, or were implicitly defined. For instance, in Harris et al., a study of belief and disbelief, participants were asked to rate phrases as "true" or "false" while their brain activation was measured with fMRI. Because this is how the authors chose to operationalize belief and disbelief, we can surmise that they defined those values, implicitly, as being akin to truth and falsity (as opposed to some other definition of belief emphasizing faith, trust, or confidence). While our survey was not exhaustive, it does suggest that rhetoricians seeking to incorporate neuroscience findings must do considerable work to unpack the assumptions underlying any single study, to put those in the context of other studies, and then to compare neuroscience understandings with those common to our own field. This preliminary survey suggests that we need to be careful not to assume that terms like "reason" or "emotion" have stable definitions, that they are defined in the same way across studies, or that they necessarily align with the preferred rhetorical definitions.

Empathy and Neurological Difference

As we have shown, neuroscience research is replete with methodological and terminological variability, so that writers of research articles must make careful choices about the terms and methods they describe. By drawing on rhetoric of science studies, readers of these articles can examine research articles carefully, considering alternative interpretations and the broader cultural debates in which such articles participate (Bazerman; Berkenkotter and Huckin; Myers; Schryer; Swales). The example we consider here is an article titled "Neural Mechanisms of Empathy in Adolescents with Autism Spectrum Disorder and Their Fathers," published by Ellen Greimel et al. in a 2010 issue of *NeuroImage*. We chose this article because it deals with a topic of great cultural interest at the moment, one suited to the emphasis of this special issue on neurological difference: autism. Not only is autism a highly debated topic in popular spheres, but, as a communicative disorder, it is sometimes posited as a kind of touchstone against which rhetorical ability can be measured (see, for instance, Oakley 102).

Formerly seen as a rare disorder, autism diagnosis rates have risen dramatically over the last twenty years, with current prevalence estimates at 1 in 110, according to the Center for Disease Control. In the *Diagnostic and Statistical Manual* (DSM-IV) of the American Psychological Association, autistic disorder (sometimes called Autistic Spectrum Disorder, or ASD), is defined in part by a list of impairments in communication and social interaction, combined with repetitive and stereotyped behavior.[6] This increase in diagnosis has led to many cultural developments: the rising influence of parent organizations arguing for biomedical treatment options; the increasing presence of autistic characters in television and film;[7] and the growing self-advocacy movement among autistic individuals who seek a greater voice in shaping directions for research and advocacy (O'Neil; Solomon; Sinclair). Scientific articles about autism necessarily participate in this broader cultural milieu. New findings about autism tend to be widely reported in the media, especially when they suggest either anatomical or genetic differences that may explain the behavioral criteria that distinguish autism.

One of these broader cultural trends is the position of ASD as a male disorder. The first thing to notice from the title of Griemel's study is that the study focused on adolescents and their *fathers*. From the abstract, we learn that the study examined high-functioning boys with a diagnosis of ASD. While the writers do not remark on their choice of male subjects, from a rhetorical standpoint these facts situate the article within a broader cultural depiction of autism as a disorder affecting males. Studies suggest that boys are four times more likely than girls to receive a diagnosis of ASD, a fact that has led some researchers to posit that autism is a disorder of the "extreme male brain" (Baron-Cohen *The Essential Difference*; Baron-Cohen "The Extreme-Male-Brain"). In this theory, ASD simply represents an exaggeration of qualities taken to be typically male. In Baron-Cohen's theory, brains tend to be either "systematizing" or "empathizing."

---

[6]At the time of writing, The American Psychiatric Association (APA) was considering proposed changes to the criteria for autism for the next edition of the *Diagnostic and Statistical Manual*, DSM-5. Previously, there were separate diagnostic categories for Asperger's Syndrome and Pervasive Developmental Disorder (PDD), to variants of autism. According to the APA, the new category would help to simplify diagnosis, since deciding where to draw the lines between sub-categories was akin to trying to "cleave meatloaf at the joints." See American Psychiatric Association..

[7]Recent examples include *Mozart and the Whale* (2005), *Adam* (2009), and the HBO biopic *Temple Grandin* (2010).

Women tend to score higher on tests of empathizing, while men tend to score higher on tests of systematizing. Nonetheless, Baron-Cohen insists that it is *brains* that are male (systemizing) or female (empathizing), not necessarily the bodies in which those brains exist.[8] At any rate, individuals with ASD, according to Baron-Cohen, get exceedingly high scores on systemizing tests. In fact, Greimel et al. used Baron-Cohen's survey of systematizing and empathizing tendencies, called the "Autistic Quotient," or AQ, to determine whether the fathers in the study possessed autistic qualities.

Rhetoric researchers might be interested in examining this broader debate about gender and autism and how it is inflected in a particular article. By focusing on male subjects, Griemel et al. subtly appeal to the dominant depiction of the disorder as fundamentally male—a depiction that also plays on the ever-popular suggestion that male and female brains are fundamentally different (Condit). This is not necessarily a shortcoming of Greimel et al.'s paper. After all, it is quite commonplace to constrain one's sample size by looking only at one sex. Recently, though, some researchers have suggested that girls and women with ASD are underdiagnosed, that the definition of the disorder itself overlooks how ASD may present in females differently (Koenig and Tsatsanis). Scientific articles like the one by Griemel et al. participate in this gendering of autism as a male disorder, a process that draws on cultural discourses about masculinity, technology, and geekiness.

By focusing on empathy, the authors of this study make a rhetorical, as well as a scientific, choice, framing their article as an intervention into that particular theory of autism's etiology. Autism presents interesting questions for neuroscientists who seek to identify differences in brain structure and function between people with and without autism. One of these proposed differences is a lack of empathy, often called mindblindness, in individuals with autism (Baron-Cohen *Mindblindness*; Happé). Accordingly, studies of empathy constitute a large proportion of autism research studies in psychology or neuroscience. Drawing on Baron-Cohen's work, Griemel et al. open by identifying "difficulties inferring their own and other persons' mental states" as among the core deficits of autism (1055). While the writers present this as a statement of fact, there are competing theories, such as the intense world hypothesis (Tager-Flusberg) or weak central coherence theory (Frith and Happé) which are not mentioned in this article. Further, the term empathy does not appear in the APA's diagnostic criteria for autism; the closest terminology in that text refers to difficulty with social reciprocity. Empathy may be an attractive concept to neuroscience researchers interested in autism because it can be operationalized in an fMRI study via quizzes or images, and

---

[8]Women can possess "male" brains, or men "female" brains, depending on how the individual scores on a test of systematizing versus empathizing, a fact that calls into question the use of the terms male and female to describe these brains in the first place.

because it has been studied in non-autistic individuals. In contrast, social reciprocity may seem fuzzier or more difficult to operationalize, and therefore more difficult to justify in a research article.[9] Rhetorical scholars should pay careful attention to how and why scientists choose specific concepts to test, how they are defined, and whether they may (or may not) apply to rhetorical concepts.

Rhetoric scholars should also pay close attention to the terminology used to describe research findings. In their study, Greimel et al. draw on genetic explanations for autism, suggesting that "[s]imilarities in neurocognitive and behavioural profiles [between individuals with ASD and their family members] strongly suggest a common biological substrate underlying these disturbances. Thus, exploring the neural underpinnings of altered social cognition in persons with ASD and their first-degree relatives might be a valuable approach to identifying familial influences on autistic pathology" (1055–1056). Here, the writers first suggest a "common biological substrate" and then replace that term, in the second sentence, with "neural underpinnings," a move that concretizes their suggestion that there may be identifiable neurological similarities between the boys with ASD and their fathers. The writers suggest that their results indicate "that FG [fusiform gyrus] dysfunction in the context of empathy constitutes a fundamental neurobiological deviation in ASD" (1062). The transformation is subtle, but what are understood to be neural *correlates* of empathy become located in the fusiform gyrus (FG), a particular site in the brain, which then becomes (potentially) a concrete, physical predictor of ASD. The term *neural correlates* is particularly slippery in this way— while it suggests correlation, not causation, the noun phrase *neural correlates* makes the phenomenon seem more concrete. To a non-scientist, especially, neural correlates may easily be confused with "neural substrates" or something similarly tangible. It is easy to overlook the fact that the researchers are mapping BOLD activity, a proxy for neurological function, to behavior under particular experimental circumstances. Rhetoric scholars, then, should pay close attention to terms such as neural correlates, neural substrates, and the like, being sure to tease out what these terms mean and the potential suasory impact of such terms.

With regards to the methodology used, Asperger's syndrome serves an interesting rhetorical and methodological function. It is also notable that the authors studied adolescent boys diagnosed with Asperger's syndrome, usually considered a high-functioning variant of autism, but one that is currently listed as a separate disorder in the DSM-IV. The writers posit that Asperger's may serve as an

---

[9]For instance, the "intense world" hypothesis suggests that ASD stems from a hyperactive, hypersensitive brain, producing exaggerated (and confusing) reactions to sensory input (see Markram et al. 19). Autistic individuals often protest the "lack of empathy" or "mind-blindness" characterization. One autistic person writes: "sometimes doctors describe autistics as though they are emotionless automatons. This is far from the truth, especially as many autistics have parents or close relatives who have bipolar disorder. You can't get more emotional than bipolar disorder. I feel things very deeply. A lack of empathy isn't central to autism, it's just a feature of the social withdrawal." See Alien Robot Girl.

appropriate analogy to other forms of autism: "One way to overcome the barriers associated with such complexity [in autistic disorders] is to examine qualitatively similar but milder phenotypes in relatives of affected individuals" (1056). The rhetorical figure at hand is the *incrementum*, which Jeanne Fahnestock suggests orders subjects who presumably share some kind of attribute to differing degrees (*Rhetorical Figures in Science* 95). The notion that individuals with classic autism and with Asperger's syndrome exist on a spectrum, or *incrementum*, implies that they differ in degree of impairment, but have the same underlying biological condition. It is this figure that grounds studies such as this one, which allow boys with Asperger's to stand in for all individuals with autism.

The notion of incrementum can help to explain the language that writers use to describe autism and Asperger's. While they portray Asperger's as a "milder pheno-type" of autism, they nevertheless described the test group as "affected indivi-duals" (1062), while the control group was called "healthy adolescents" (1063) or "healthy controls" (1063). Individuals with ASD were characterized as having "*aberrant* neural face and mirroring mechanisms" (1055, emphasis ours) and "socioemotional *impairments*" (1063, emphasis ours). This terminology, common to scientific articles about autism, also constitutes a rhetorical choice among available terms. By choosing this language, the writers position their work clearly within a scientific conversation surrounding the deficits apparent in autism. A different choice of language might signal a different approach. For instance, individuals who argue for neurodiversity, or the notion that neurological differ-ences are at least partially culturally produced, might use terms such as *difference*, *condition* (as opposed to *disorder*), and *acceptance* rather than *therapy* or *cure*. Meanwhile, parents who advocate biomedical treatments for autism tend to use terms connoting *disease* and *devastation*, on the one hand, and *cure* or *recovery*, on the other, to argue for their case. For rhetoric scholars, the point may not be to weigh in on this debate, but to pay attention to the kinds of language used to describe a neurological difference such as autism, and to the different meanings they carry in different contexts.

The implications of any scientific article, which usually appear in the discussion section, are key considerations. In scientific articles on autism, diagnosis and gen-etics tend to appear in discussion sections, serving as commonplaces to help wri-ters address the so-what question. The upshot of Griemel et al.'s implications is that autism might be corrected through medical intervention, particularly early diagnosis and genetic identification. Griemel et al. suggest that "Illuminating aber-rancies such as reduced activation of the amygdala and the FG in persons present-ing with mild autistic traits might prove beneficial for the identification of neurobiological endophenotypes of ASD and may provide future directions for molecular genetic studies" (1063). These commonplaces are disputed in other circles, such as in the neurodiversity movement, where they are interpreted as por-tending the possibility of fetal screening and selective abortion of fetuses identified as "autistic." Meanwhile, parents who write about autism often embrace early diagnosis and screening techniques, since they may help them to argue for appropriate therapies and resources. For rhetoric scholars, then, it is key to consider how these topoi function as rhetorical choices, and how those topoi might be interpreted in other contexts.

One might be inclined to conclude from Griemel et al.'s study that the fusiform gyrus (FG) can verify the fundamentally human capacity for empathy, or the lack thereof in autistic individuals, and hence rhetoric. As we mentioned earlier, empathy underlies a number of rhetorical theories, including those of Quintilian and Burke. Indeed, Dennis Lynch argues that "The concept and practice of empathy insinuates itself into most modem rhetorical theories, under one guise or another" (6). It might be tempting, then, to use the case of autism as a kind of test case or touchstone against which "normal" human rhetorical capacities might be measured. Using empathy as a marker of rhetorical potential might seem to exclude individuals with autism from human rhetorical capacity on almost every level, a fact that can be ethically objectionable. A more responsible move, then, might be to question whether it is ethical for rhetoricians to assume a lack of empathy in other humans, or to consider whether rhetorical theories should be revised in order to better account for the full range of human rhetorical capacities, including those with neurological differences.

Any research article is situated both with relation to a scientific conversation and a broader cultural one. In the case of neurological differences, these contexts are increasingly convergent, in that non-scientists are gaining a voice in research decisions about autism, bipolar disorder, depression, and the like. However, readers must be quite familiar with such debates, both within scientific communities and outside of them, in order to understand the rhetorical choices, as well as elisions, within a given article.

Rhetorical Considerations: Neuroscience Findings in the Popular Media

Given the complexities of scientific texts, rhetoric scholars might be drawn to popular texts about neuroscience, since they provide accessible overviews of current findings. In general, though, popular science reports often repackage scientific findings by drawing on topoi such as application or wonder (Fahnestock "Accommodating"). In their study of popular news reports about neuroscience, in particular, Erik Racine, Bar-Ilan Ofek, and Judy Illes, categorize claims as falling into three types, which they call neuro-realism, neuro-essentialism, and neuro-policy. Readers should be aware of these three commonplaces, how they work on audiences, and how they might relate to the scientific reports themselves, rather than taking them at face value. Moreover, readers might also look for these tendencies in scientific articles, where they may appear in the discussion section as a way to signal the importance of a given research study.

As described by Racine, Ofek, and Illes, neuro-realism occurs when "coverage of fMRI investigations can make a phenomenon uncritically real, objective or effective in the eyes of the public" (160), or when reports invalidate or validate our ordinary understanding of the world. We would suggest that neuro-realism can occur in popularization of all kinds of neuroscience research, not just those that report on fMRI research. Rhetorically, neuro-realism operates through metaphors that work to spatially locate specific functions in the brain. One example of neuro-realism is this headline from *New Scientist*: "Emotional speech leaves 'signature' on the brain" (Thomson). In this study, scientists examined patterns of brain activation in 22 individuals who listened to a single sentence, read with different emotional inflections. In the article, Thomson suggests that the scientists observed "signatures," a term that implies that the results in question somehow left a mark on the brain, rather than interpreting them as momentary patterns of activation. The usage reflects the underlying metaphor of the brain as text, inscribed by sensory experiences. A correlate of this metaphor tends to be the suggestion that scientists can therefore "read minds" in a popular sense, as though scientists could literally read a transcript of someone's thoughts rather than *interpret* visual images or data. Such usage, along with references to regions of the brain such as a "emotion center," "neural architecture," or "god spot," also involve spatial metaphors, which, like textual metaphors, seek to fix brain functions in particular spaces.

The second tendency, neuro-essentialism, refers to "how fMRI research can be depicted as equating subjectivity and personal identity to the brain" (160). The key rhetorical figure for neuro-essentialism might be a double synechdoche, wherein both the brain and the quality to be measured stand in for a complex of biological and cultural factors. An example of might be this claim from a MSNBC report: "Two new brain-imaging studies describing the origins of empathy and how placebos work provide insights into the nature of pain, the mind-body connection and what it means to be human" (Kane). The "brain" stands in for the complex network of neurons, blood flow, bodily actions, and cognitive processes that might actually make up something like pain or the "mind-body connection." Once the brain takes over for this complex, it can be given tasks like "handling love and pain" or telling us "what it means to be human." In this way, the brain represents the *essence* of human experiences (love, pain), or even of humanness itself. Such reports often anthropomorphize the brain, making it an active agent, as in the headline for the Kane article, "How your brain handles love and pain."

Finally, neuro-policy refers to "attempts to use fMRI results to promote political and personal agendas" (Racine, Bar-Ilan, and Illes 161). Often, neuro-policy arguments rely on weak analogies that extend the initial research findings far beyond their original contexts. For example, a recent report in *Popular Science* suggested that a new study showing neural correlates of pain in 16 men undergoing oral surgery held implications for animal rights: "applications of this technology for fields beyond medicine, such as animal rights, may prove more transformative than any medical use. Using the fMRI on animals could quantify the pain levels of veterinary and slaughter procedures, potentially changing the way we both heal and kill animals" (Fox). Here, the writer extends the research findings beyond their immediate context (research on humans undergoing oral surgery) to a very different context—animals being treated by a veterinarian or slaughtered for food. This is not to say that such an application might not be possible, or warranted, but these kinds of statements tend to minimize the time, effort, and technological innovation required for these applications. A weak analogy can also occur when writers extend animal studies to humans, since the biological and social factors influencing human cognition vary from, say, rats. Whether or not the comparison is apt depends on the situation.

Given these tendencies in popular and scientific articles, researchers in rhetoric should carefully question the interpretations writers give for neuroscience findings. More generally, both popular and scientific texts take advantage of the persuasiveness of visual images of the brain (McCabe and Castel; Johnson; Beaulieu "Images"; Weisberg et al.), sometimes leading audiences to grant greater credibility to scientific claims than they might otherwise. These images, nearly ubiquitous in popular reports, act as the warrant for the scientific claims, offering what Ann Beaulieu calls "a concrete unit of scientific knowledge" in an attractive, visual form ("Images" 54). Rhetoricians seeking to rely on popularized accounts of neuroscience need to carefully read such texts with a rhetorical lens, considering how authors frame their research, what arguments they make, and what other viewpoints might exist in the field.

Guidelines and Future Directions

Neuroscience research holds tantalizing possibilities for rhetorical scholarship; however, there exists a deep divide in how rhetoricians and neuroscientists communicate. Paying attention to the fundamental differences in discourse between these communities will help rhetoric scholars to work with neuroscience findings in a responsible manner. To conclude, we offer here some guidelines for scholars in rhetoric who plan to use neuroscience research in their work.

First, as we showed in the initial section, it is important to understand the methodological assumptions and debates driving neuroscience research. While neuroimaging technologies provide powerful tools, their interpretation is also guided by complex, ongoing arguments about specific methodological practices. In fact, untangling how methodological assumptions influence neuroscientific analyses sometimes reveals that these assumptions may predetermine results to some extent. By focusing on the topoi of *accuracy, precision,* and *bias,* we showed how neuroscientists negotiate the empirical parameters out of which claims about the brain can be made. Since these parameters are not yet settled, it is important for outside readers to be careful about applying scientific results to new contexts, such as rhetorical ones.

Second, rhetoric scholars should carefully compare their own terms and definitions with those used in neuroscience fields. As we showed in our second section, the neuroscience concepts of emotion, reason, and empathy might seem to match rhetorical concepts of pathos, logos, and identification; yet this assumption would be false in many cases. In order to draw conclusions from neuroscientific studies, rhetoric scholars will either have to make judgments about what qualifies as a close enough operationalization of a given concept or work directly with neuroscientists to operationalize our own concepts. Rhetoric scholars might work with neuroscientists to empirically test rhetorical effectiveness of, say, campaign speeches. However, both of us are skeptical of this approach, which could easily fall into the traps of neurorealism or neuroessentialism.

Third, scholars should draw on the insights of rhetoric of science scholarship when examining any particular scientific article. Rather than simply extracting the findings from an abstract, we should be careful to consider the framework the writers create for their data, asking what other scientific frameworks or explanations might be possible, and how those frameworks relate to broader debates. Such an approach requires at least some familiarity with the conceptual or methodological debates going on within a given field of study. For neuroscientists, evaluating a single article depends on the ability to fit that article within a body of converging evidence. For scholars doing interdisciplinary work, this poses a significant challenge. Ideally, collaborative work with neuroscience researchers could provide an avenue for rhetoric scholars to use neuroscience insights responsibly in their own work. In lieu of a direct collaboration, rhetoric scholars will need to do significant outside reading in order to situate a given research finding within its discourse community.

Finally, rhetoric scholars should be wary of repeating (or making their own) claims that fall into the trap of neurorealism, neuroessentialism, or neuropolicy. While we locate these pitfalls in popular accounts, they may also be tendencies in the cross-disciplinary endeavor we are calling neurorhetorics. For instance, we might be apt to argue that a given rhetorical concept (pathos, ethos, identification, or what have you) can be proven to exist due to neuroimaging studies—an instance of neurorealism. Or, we might lean toward neuroessentialism by claiming that brain scan studies attest to different types of brains—the "pathos-driven brain" or the "logos-driven brain"—based on how individuals respond to emotional versus logical kinds of arguments. The third pitfall, neuropolicy, may be especially likely to ensnare scholars interested in rhetorical production. Suggesting that brain studies offer proof of the effectiveness of a specific method of instruction will often fall into the trap of neuropolicy. Clearly, the classroom is a much more complex place than can be simulated inside an fMRI machine, so we should be wary of weak analogies that seek to offer scientific proof of the effectiveness of any given pedagogical method.

Given these caveats, we return to our initial claim that neurorhetorics research should involve both careful rhetorical analysis of neuroscience arguments as well as consideration of how neuroscience can inform rhetorical theory and practice. Accordingly, we suggest the following directions for future research.

We may consider how scientific research about the brain is used to support arguments in all kinds of venues (political, legal, literary, medical, and so on). In this issue, for instance, Katie Guest Pryal, John P. Jackson, and Jenell Johnson each examine the rhetorical controversies that often surround attempts to define (or exclude) individuals on the basis of cognitive functioning or difference. Rhetoricians might also consider why neuroscientific explanations and images hold particular sway over audiences, an issue neuroscientists have themselves been debating. In one recent study, researchers David McCabe and Alan Castel found that people ranked descriptions of brain research as more credible if it included images of brain scans. In another study, researchers found that even including the words "brain scans indicate" increased readers' confidence in explanations of brain phenomena, leading researchers to question the "seductive allure" that neuroscience seems to hold (Weisberg et al.). Rhetoricians might help us to understand why neuroscience findings seem to hold this allure, and to what effect.

While scholars have shown that popular news accounts tend to overemphasize or decontextualize neuroscience research findings (McCabe and Castel; Weisberg et al.), we know of no studies that question whether neuroscience research articles themselves reflect popular science preoccupations. In other words, do publication and funding pressures encourage authors to frame their research in ways that will lead to attractive headlines? A related project might consider how rhetorical theory can account for the persuasiveness of neuroimaging and neuroscience findings noted by McCabe and Castel and Weisberg et al.

Applications of neuroscience research to legal contexts should also interest rhetorical scholars, since forensics have always been part of rhetoric's domain. Brain data are already being offered as evidence in trials, but we contend that such data must be interpreted by expert witnesses.[10] For rhetoric scholars this means that expert ethos becomes a key issue—who is trusted to make these judgments in court (and in other venues)? How do legal and scientific arguments coincide in these cases? The legal venue is just one of many in which neurological difference is produced, identified, and realized in specific brains. As the articles in this issue demonstrate, the production of neurological difference never happens exclusively within scientific realms, but it nonetheless draws on neuroscience evidence for its power.

Finally, we hope more researchers in both rhetoric and neuroscience will undertake collaborative, interdisciplinary research. As we have shown, these two fields do have much to say to one another. While great differences in research methodologies, foundational concepts, discourse practices, and publication venues

---

[10]See Feigenson for a discussion of the admissibility and persuasiveness of fMRI data as courtroom evidence.

exist between these fields, we hope that the two-sided approach proposed in this article can help rhetoric scholars to use neuroscience insights in a responsible manner to yield productive insights into rhetoric while minimizing potential pitfalls of interdisciplinary work. We have highlighted a number of strategies here, such as carefully considering neuroscience research methods, comparing neuroscientific with rhetorical understandings of similar terms, or grounding any borrowings in a broader understanding of rhetorical and scientific debates surrounding neurological difference. Given the great interest and importance of these disciplines we encourage future research projects that have the potential to produce further productive interchanges between these fields.

Acknowledgments

The authors thank Scott Huettel, Nico Boehler, Debra Hawhee, and Katie Rose Guest Pryal for their valuable comments on this article.

References

Akitsuki, Yuko, and Jean Decety. "Social Context and Perceived Agency Affects Empathy for Pain: An Event-Related fMRI Investigation." *NeuroImage* 47.2 (2009): 722–734. Print.

Alien Robot Girl. Comment on "Do Supercharged Brains Give Rise to Autism?" *Plant Poisons and Rotten Stuff Blog*, 2008. Web.

Arthos, John. "Locating the Instability of the Topic Places: Rhetoric, Phronesis and Neurobiology." *Communication Quarterly* 48.2 (2000): 272–292. Print.

Azim, Eiman, et al. "Sex Differences in Brain Activation Elicited by Humor." *Proceedings of the National Academy of Sciences of the United States of America* 102.45 (2005): 16496–164501. Print.

Bandettini, Peter A. "Seven Topics in Functional Magnetic Resonance Imaging." *Journal of Integrative Neuroscience* 8.3 (2009): 371–403. Print.

Baron-Cohen, Simon. *The Essential Difference: The Truth About the Male and Female Brain.* New York: Basic Books, 2003. Print.

———. "The Extreme-Male-Brain Theory of Autism." *Trends in Cognitive Science* 6.6 (2002): 248–254. Print.

———. *Mindblindness: An Essay on Autism and Theory of Mind.* Cambridge: MIT Press, 1997. Print.

Bazerman, Charles. *Shaping Written Knowledge.* Madison: University of Wisconsin Press, 1988. Print.

Beaulieu, Anne. "From Brainbank to Database: The Informational Turn in the Study of the Brain." *Studies in History and Philosophy of Biology & Biomedical Science* 35 (2004): 367–390. Print.

———. "Images Are Not the (Only) Truth: Brain Mapping, Visual Knowledge, and Iconoclasm." *Science, Technology, & Human Values* 27 (2002): 53–86. Print.

Berkenkotter, Carol, and Thomas N. Huckin. "You Are What You Cite: Novelty and Intertextuality in a Biologist's Experimental Article." *Professional Communication: The Social Perspective.* Eds. Blyler, Nancy Roundy & Charlotte Thralls. Newbury Park, CA: Sage, 1993. 109–127. Print.

Burke, Kenneth. *A Rhetoric of Motives.* Berkeley: University of California Press, 1969. Print.

Canessa, Nichola, et al. "The Effect of Social Content on Deductive Reasoning: An fMRI Study." *Human Brain Mapping* 26 (2005): 30–43. Print.

Carter, R. M., et al. "Activation in the VTA and Nucleus Accumbens Increases in Anticipation of Both Gains and Losses." *Frontiers in Behavioral Neuroscience* 3 (2009): 21. Print.

Condit, Celeste. "How Bad Science Stays That Way: Of Brain Sex, Demarcation, and the Status of Truth in the Rhetoric of Science." *Rhetoric Society Quarterly* 26 (1996): 83–109. Print.

Fahnestock, Jeanne. "Accommodating Science: The Rhetorical Life of Scientific Facts." *Written Communication* 15.3 (1998): 330–350. Print.

———. *Rhetorical Figures in Science.* New York: Oxford University Press, 1999. Print.

Feigenson, Neil. "Brain Imaging and Courtroom Evidence: On the Admissibility and Persuasiveness of fMRI." *International Journal of Law in Context* 2.3 (2006): 233–255. Print.

Ferstl, Evelyn C., and D. Yves von Cramon. "Time, Space and Emotion: fMRI Reveals Content-Specific Activation During Text Comprehension." *Neuroscience Letters* 427 (2007): 159–164. Print.

Flower, Linda. "Cognition, Context, and Theory Building." *College Composition and Communication* 40.3 (1989): 282–311. Print.

Fox, Stuart. "New Brain Scan Quantifies the Formerly Subjective Feeling of Pain." *Popular Science* (2010). Web.

Friston, K., and R. Henson. "A Commentary on Divide and Conquer; a Defense of Functional Localisers." *NeuroImage* 30 (2006): 1097–1099. Print.

Friston, K. J., et al. "A Critique of Functional Localisers." *NeuroImage* 30.4 (2006): 1077–1087. Print.

Frith, U., and F Happé. "Autism: Beyond 'Theory of Mind.'" *Cognition* 50 (1994): 115–132. Print.

Greene, Joshua D., et al. "The Neural Bases of Cognitive Conflict and Control in Moral Judgment." *Neuron* 44.2 (2004): 389–400. Print.

Greimel, Ellen, et al. "Neural Mechanisms of Empathy in Adolescents with Autism Spectrum Disorderand Their Fathers." *NeuroImage* 49 (2010): 1055–1065. Print.

Gross, Daniel M. *The Secret History of Emotion: From Aristotle's 'Rhetoric' to Modern Brain Science.* Chicago: The University of Chicago Press, 2006. Print.

Happé, F. G. "An Advanced Test of Theory of Mind: Understanding of Story Characters' Thoughts and Feelings by Able Autistic, Mentally Handicapped, and Normal Children and Adults." *Journal of Autism and Developmental Disorders* 24.2 (1994): 129–154. Print.

Harris, Sam, Sameer A. Sheth, and Mark S. Cohen. "Functional Neuroimaging of Belief, Disbelief, and Uncertainty." *Annals of Neurology* 63 (2008): 141–147. Print.

Huettel, Scott A., Gregory McCarthy, and Allen W. Song, eds. *Functional Magnetic Resonance Imaging.* Sunderland, MA: Sinauer Associates, 2004. Print.

Johnson, Davi. "'How Do You Know Unless You Look?': Brain Imaging, Biopower and Practical Neuroscience." *Journal of Medical Humanities* 29 (2008): 147–161. Print.

Kane, Daniel. "How Your Brain Handles Love and Pain." *MSNBC.* February 8 (2004). Web.

Kelland, Kate. "Vegetative Patient 'Talks' Using Brain Waves." *Reuters* 3 Febraury 2010. Web.

Kennerley, S. W., et al. "Neurons in the Frontal Lobe Encode the Value of Multiple Decision Variables." *Journal of Cognitive Neuroscience* 21.6 (2009): 1162–1178. Print.

Koch, K., et al. "Gender Differences in the Cognitive Control of Emotion: An fMRI Study." *Neuropsychologia* 20 (2007): 2744–2754. Print.

Koenig, Kathleen, and Katherine D. Tsatsanis. "Pervasive Developmental Disorders in Girls." *Handbook of Behavioral and Emotional Problems in Girls.* Eds. Debora J. Bell, Sharon L. Foster, and Eric J. Mash. New York: Kluwer Academic/Plenum Publishers, 2005. Print.

Kompus, K., et al. "Distinct Control Networks for Cognition and Emotion in the Prefrontal Cortex." *Neuroscience Letters* 467.2 (2009): 76–80. Print.

Krämer, Ulrike M., et al. "Emotional and Cognitive Aspects of Empathy and Their Relation to Social Cognition—An fMRI-Study." *Brain Research* 1311 (2010): 110–120. Print.

Labar, K. S. "Beyond Fear Emotional Memory Mechanisms in the Human Brain." *Current Directions in Psychological Science* 16.4 (2007): 173–177. Print.

Lewiecki-Wilson, Cynthia. "Rethinking Rhetoric through Mental Disabilities." *Rhetoric Review* 22.2 (2003): 156–167. Print.

Littlefield, Melissa. "Constructing the Organ of Deceit: The Rhetoric of Fmri and Brain Fingerprinting in Post-9/11 America." *Science, Technology, & Human Values* 34 (2009): 365–392. Print.

Logothetis, N. K., et al. "Neurophysiological Investigation of the Basis of the fMRI Signal." *Nature* 412.6843 (2001): 150–157. Print.

Logothetis, N. K., and B. A. Wandell. "Interpreting the Bold Signal." *Annual Review of Physiology* 66 (2004): 735–769. Print.

Lu, Shengfu, et al. "Recruitment of the Pre-Motor Area in Human Inductive Reasoning: An fMRI Study." *Cognitive Systems Research* 11.1 (2010): 74–80. Print.

Lynch, Dennis A. "Rhetorics of Proximity: Empathy in Temple Grandin and Cornel West." *Rhetoric Society Quarterly* 28.1 (1998): 5–23. Print.

Lynch, Michael E. "Science and the Transformation of the Animal Body into a Scientific Object: Laboratory Culture and Ritual Practice in the Neurosciences." *Social Studies of Science* 18.2 (1988): 265–289. Print.

Mak, Amanda K. Y., Zhi-guo Hu, John X. Zhang, Zhuang-wei Xiao, and Tatia M. C. Lee. "Neural Correlates of Regulation of Positive and Negative Emotions: An fMRI Study." *Neuroscience Letters* 457 (2009): 101–106. Print.

Markram, Henry, Tania Rinaldi, and Kamila Markram. "The Intense World Syndrome—An Alternative Hypothesis for Autism." *Frontiers in Neuroscience* 1 (2007): 19. Print.

McCabe, David P., and Alan D. Castel. "Seeing is Believing: The Effect of Brain Images on Judgments of Scientific Reasoning." *Cognition* 107 (2008): 343–352. Print.

McRae, Kateri, et al. "Gender Differences in Emotion Regulation: An Fmri Study of Cognitive Reappraisal." *Group Processes & Intergroup Relations* 11.2 (2008): 143–162. Print.

Mooney, Chris, and Sheril Kirshenbaum. *Unscientific America: How Scientific Illiteracy Threatens Our Future.* New York: Basic Books, 2009. Print.

Myers, Greg. *Writing Biology.* Madison: The University of Wisconsin Press, 1990. Print.

O'Neil, Sara. "The Meaning of Autism: Beyond Disorder." *Disability & Society* 23.7 (2008): 787–799. Print.

Oakley, Todd V. "The Human Rhetorical Potential." *Written Communication* 16 (1999): 93–128. Print.

Ochsner, Kevin N., et al. "Reflecting Upon Feelings: An fMRI Study of Neural Systems Supporting the Attribution of Emotion to Self and Other." *Journal of Cognitive Neuroscience* 16.10 (2004): 1746–1772. Print.

———. "Rethinking Feelings: An fMRI Study of the Cognitive Regulation of Emotion." *Journal of Cognitive Neuroscience* 14.8 (2002): 1215–1229. Print.

Padoa-Schioppa, C., and J. A. Assad. "Neurons in the Orbitofrontal Cortex Encode Economic Value." *Nature* 441.7090 (2006): 223–226. Print.

Platt, M. L., and P. W. Glimcher. "Neural Correlates of Decision Variables in Parietal Cortex." *Nature* 400.6741 (1999): 233–238. Print.

Poldrack, R. A., et al. "Guidelines for Reporting an fMRI Study." *Neuroimage* 40.2 (2008): 409–414. Print.

Poldrack, Russell A. "Region of Interest Analysis for fMRI." *Social Cognitive and Affective Neuroscience* 2.1 (2007): 67–70. Print.

Prelli, Lawrence J. *A Rhetoric of Science: Inventing Scientific Discourse.* Columbia, SC: University of South Carolina Press, 1989. Print.

Quintilian. Trans. H.E. Butler The Loeb Classical Library. *The Instituto Oratoria of Quintilian*. Eds. T. E. Page, E. Capps, and W. H. D. Rouse. Cambridge, MA: Harvard University Press, 1936. Print.

Racine, Eric, Ofek Bar-Ilan, and Judy Illes. "fMRI in the Public Eye." *Nature Reviews Neuroscience* 6 (2005): 159–164. Print.

Rangel, A. "The Computation and Comparison of Value in Goal-Directed Choice." *Neuroeconomics: Decision Making and the Brain*. Eds. P. W. Glimcher et al. New York: Academic Press, 2009. 425–439. Print.

Reuter, M., et al. "Personality and Emotion: Test of Gray's Personality Theory by Means of an fMRI Study." *Behavioral Neuroscience* 118.3 (2004): 462–469. Print.

Sander, David, et al. "Emotion and Attention Interactions in Social Cognition: Brain Regions Involved in Processing Anger Prosody." *NeuroImage* 28.4 (2005): 848–858. Print.

Saxe, R., M. Brett, and N. Kanwisher. "Divide and Conquer: A Defense of Functional Localizers." *Neuroimage* 30.4 (2006): 1088–1096; discussion 97–9. Print.

Schaich, Borg J., et al. "Consequences, Action, and Intention as Factors in Moral Judgments: An Fmri Investigation." *Journal of Cognitive Neuroscience* 18.5 (2006): 803–817. Print.

Schryer, Catherine F. "Genre Time/Space: Chronotopic Strategies in the Experimental Article." *JAC* 19 (1999): 81–89. Print.

Sinclair, Jim. "Don't Mourn for Us." *Autism Network International Newsletter* 1.3 (1993). Web.

Solomon, Andrew. "The Autism Rights Movement." *New York Magazine* 25 May 2008. Print.

Specter, Michael. *Denialism: How Irrational Thinking Hinders Scientific Progress, Harms the Planet, and Threatens Our Lives*. New York: Penguin, 2009. Print.

Swales, John. *Genre Analysis: English in Academic and Research Settings*. Cambridge: Cambridge University Press, 1990. Print.

Tager-Flusberg, Helen. "Evaluating the Theory-of-Mind Hypothesis of Autism." *Current Directions in Pscyhological Science* 16.6 (2007): 311–315. Print.

Thomson, Helen. "Emotional Speech Leaves 'Signature' on the Brain." *New Scientist* 2009. Web.

Trautmann, SinaAlexa, Thorsten Fehr, and Manfred Herrmann. "Emotions in Motion: Dynamic Compared to Static Facial Expressions of Disgust and Happiness Reveal More Widespread Emotion-Specific Activations." *Brain Research* 1284 (2009): 100–115. Print.

Turner, Mark. "The Cognitive Study of Art, Language, and Literature." *Poetics Today* 23.1 (2002): 9–20. Print.

Weisberg, Deena Skolnick, et al. "The Seductive Allure of Neuroscience Explanations." *Journal of Cognitive Neuroscience* 20.3 (2008): 470–477. Print.

Zimmerman, Frederick J., Dimitri A. Christakis, and Andrew N. Meltzoff. "Associations between Media Viewing and Language Development in Children under Age 2 Years." *The Journal of Pediatrics* 151.4 (2007): 364–368. Print.

Appendix

Article	Definition (y/n)	How defined?	How operationalized?
Kompus, Kristiina, et al.	No	n/a	Subjects presented with neutral or emotional faces; emotional faces were negative (fear or anger)
Krämer, Ulrike, et al.	Yes	Encyclopedia Britannica definition of empathy (p. 110) + Decety et al.'s list of components (3 of them)	Chose one of Decety et al.'s components to operationalize using faceless cartoons of "emotionally charged" vs. "emotionally neutral" situations that they had created
Mak, Amanda K.Y., et al.	No	Defines "emotion regulation" as dealing with 'socially appropriate' behavior.	Used "emotional pictures" from the International Emotion Picture System plus extra pictures from popular media
Harris, Sam, et al.	No	Implicit – belief equated with truth, disbelief equated with falsity	Used fMRI to study the brains of 14 adults while they judged written statements to be "true" (belief), "false" (disbelief), or "undecidable" (uncertainty)
Ferstl, Evelyn C. and D. Yves von Cramon.	No	n/a	Twenty participants read two sentence stories half of which contained inconsistencies concerning emotional, temporal or spatial information
Koch K., et al.	No	n/a	"Induced negative emotion by means of negative olfactory stimulation (with rotten yeast)" (2745), which is "an effective standardized and validated method of mood induction" (2745).
Schaich Borg J., et al.	Yes	"For our purposes, 'emotions' are immediate valenced reactions that may or may not be conscious. We will focus on emotions in the form of negative affect. In contrast, 'reason' is neither valenced nor immediate insofar as reasoning need not incline us toward any specific feeling and combines prior information with new beliefs or conclusions and usually comes in the form of cognitive manipulations…that require working memory" (803).	Presented scenarios to subjects using both "dramatic (colorful)" and "muted (noncolorful)" language.

Author			
Azim, Eiman, et al.	No	(humor)	Showed cartoons that had previously been rated "funny" or "unfunny"
Sander, David, et al.	No	No, but emphasis here is on decoding/interpreting affective *cues*.	Voices – subjects listened to meaningless utterances read in angry vs. neutral prosody.
Canessa, Nichola, et al.	Yes	"Reasoning can be defined as a combination of cognitive processes that allows us to draw inferences from a given set of information and reach conclusions that are not explicitly available, providing new knowledge. Reasoning is the central nucleus of mental activity, from text comprehension to problem solving and decision making" (930).	"In the present study, the effect of content on brain activation was investigated with functional magnetic resonance imaging (fMRI) while subjects were solving two versions of the Wason selection task, which previous behavioral studies have shown to elicit a significant content effect" (930).
Greene, Joshua D., et al.	Yes	"By "reasoning," we refer to relatively slow and deliberative processes involving abstraction and at least some introspectively accessible components" (389).	Subjects were asked to make various kinds of moral judgements (impersonal and personal) based on scenarios validated in an earlier study by Greene et al.
Reuter, M., et al.	No	n/a	"Subjects viewed pictures with sadomasochistic, erotic, disgusting, fear-inducing, and affectively neutral content. E.g. The disgusting pictures included unusual food, disgusting animals, poor hygiene, and body products such as excrement; fear-inducing pictures included scenes of animal threat, human threat, or disasters; erotic pictures contained either pictures of single naked subjects or pictures of couples in an intimate situation." (464)
Ochsner, Kevin N., et al. "Reflecting Upon Feelings"	No	n/a	"In this task, participants were presented with a series of blocks of photographic images and for each block were asked to judge either their own emotional response to each photo (pleasant, unpleasant or neutral), or to judge whether the image had been taken (indoors, outdoors, or not sure). The present study modified this paradigm through the inclusion of a third condition, which asked participants to judge the emotional response of the central character in each image (pleasant, unpleasant, or neutral)." (1748)

(Continued)

33

Appendix Continued

Article	Definition (y/n)	How defined?	How operationalized?
Ochsner, Kevin N., et al. "Rethinking Feelings"	Maybe?	"The cognitive transformation of emotional experience has been termed "reappraisal."" (1215)	"We employed two conditions: On "Attend trials," participants were asked to let themselves respond emotionally to each photo by being aware of their feelings without trying to alter them. On "Reappraise trials," participants were asked to interpret photos so that they no longer felt negative in response to them. . . . Each trial began with a 4-sec presentation of a negative or neutral photo, during which participants were instructed simply to view the stimulus on the screen" (1217).
McRae, Kateri, et al.	Sort of	"Emotion regulation can be deliberate or habitual, conscious or unconscious, and can involve changes in the magnitude, duration, or quality of one or several components of an emotional response. Emotion regulation strategies can target one's own emotions or those of another individual, at a variety of time points in the emotion generation process (Gross, 2007). Because emotion regulation is an ongoing process, the overall trajectory of an emotional response can be characterized by the effects of regulation as much as the effects of 'pure' reactivity."	"At the start of each trial, an instruction word was presented in the middle of the screen ('decrease' or 'look'; 4 seconds), a picture was presented (negative if instruction was decrease (regulation instruction), negative or neutral if instruction was look (non-regulation instruction; 8 seconds) followed by a rating period (scale from 1–4; 4 seconds) and then the word 'relax' (4 seconds). . . . Following presentation of each picture, participants were prompted to answer the question 'How negative do you feel?' on a scale from 1 to 4 (where 1 was labeled 'weak' and 4 was labeled 'strong'). (147)
Trautmann, Sina Alexa, Thorsten Fehr, and Manfred Herrmann.	No	n/a	"A set of emotional videos and video screen captures showing different facial expressions (see Fig. 4) was applied for the fMRI study. The stimuli were depicted from a stimulus data base of 40 female and 40 male non-professional actors displaying each of eight different emotional facial expressions (happiness (smiling and laughing), surprise, enjoying, fear, anger, sadness, and

34

		disgust) and neutral expressions . . . Emotional expressions of actresses were triggered by a mood induction strategy (e.g., for disgust: "imagine, you come home after two weeks of vacation but you forgot to take out the biowaste container" or happy: "imagine you meet someone unexpectedly on the street who you really like and give him a smile because you are happy to see him"). For the purposes of the present study, only female dynamic and static emotional face stimuli (N = 40; see above for detailed explanation of gender differences) displaying positive (happiness), negative (disgust), and neutral expressions were used." (111)
Akitsuki, Yuko, and Jean Decety.	Yes	The perception of pain in others can be used as a window to investigate the neurophysiological mechanisms that underpin the experience of empathy, i.e., the capacity to share and appreciate others emotional and affective states in relation to oneself (Decety, 2007; Goubert et al., 2009; Jackson et al., 2005). Empathy may be regarded as a proximate factor motivating prosocial behaviors and is crucial in the development of moral reasoning (Decety and Meyer, 2008). (722) "The task consisted of the successive presentation of animated visual images of hands and feet depicting painful and non-painful situations. Furthermore, these situations involved either an individual whose pain was caused by accident or an individual whose pain was inflicted on purpose by another person. A series of 144 stimuli were created and validated for this study. Validation of the material was conducted with a group of 222 participants (110, females) who were shown these dynamic stimuli and asked to estimate how painful these situations were and whether they believed that the pain was caused intentionally (Estabrook, 2007). Each animation consisted of three digital color pictures, which were edited to the same size (600 × 480 pixels). The durations of the first, second and third pictures were 1000 ms, 400 ms and 1000 ms respectively. These animated stimuli contained scenes of various types of painful and non-painful everyday situations" (723).
Lu, Shengfu, et al.	Yes	"Inductive reasoning is defined as the process of inferring a general rule (conclusion) by observation and analysis of specific instances (premises). Inductive reasoning is used when generating hypotheses, formulating theories and discovering relationships, and is essential for scientific discovery." (74) "The experimental tasks were adapted from a kind of intelligence test problems. The basal element of the task was a reverse triangle as shown in Figure 1. The three numbers located at three different positions may constitute a calculation rule, i.e., an equation like $Z = X + Y$. Figure 2 gave an example of the inductive reasoning task, which was assembled with three reverse triangles as mentioned above" (75).

35

The Neuroscience of Rhetoric: Identification, Mirror Neurons, and Making The Many Appear

David Gruber

In this volume, Jordynn Jack and Gregory L. Appelbaum have pointed out two reigning approaches in rhetoric to studying "all things neuro" (413). The first is "the rhetoric of neuroscience"—the "inquiry into the modes, effects, and implications of scientific discourses about the brain" (413). The second is "the neuroscience of rhetoric"—the drive for "new insights into language, persuasion, and communication from neuroscience research" (413). Together, these two approaches forge the subfield of "neurorhetoric," wherein rhetorical scholars move back and forth between analyzing work "with a rhetorical as well as a scientific lens" and seek to learn "how knowledge is established rhetorically and empirically in the field of cognitive neuroscience" (414).

The first half of neurorhetoric—the rhetoric of neuroscience—follows from work in the rhetoric of science by scholars such as Leah Ceccarelli, Alan Gross, Lisa Keränen, Carolyn Miller, and Lawrence Prelli. Interest in rhetoric of neuroscience also finds support in recent calls for rhetorical scholars to pay special attention to neuroscience publications (Fahenstock; Gibbons; Gruber et al.). And without question, the attendant research has thus far proved productive and interesting. Jack, for instance, examines the construction of autism in cognitive neuroscience by exploring the way multiple studies into autism primarily choose male participants and organize autism as a disorder of the "extreme male brain," reifying a gender bias about the constitution of males and females (421–3). Davi Johnson, to take another example, has explored how brain imagining is a "persuasive visual rhetoric" that contributes to the "self-fashioning" of "desirable selves" (147). And Michelle Gibbons has demonstrated how brain images argue differently as they move from academic journals to popular science publications (175). In short, the rhetoric of the neurosciences is starting to flourish, and rhetorical scholars will continue to explore how the language and practices of the neurosciences allow it to hold together and influence audiences.

The second half of neurorhetoric—the neuroscience of rhetoric—is pursued with less precedent from the field of rhetoric.[1] This is at least partially due to the complicated historical relationship with the sciences going back to Alan Gross' statement in the 1990s that science is "rhetoric without remainder" (6), or, if one chooses to trace the tension back through time, then Plato's division of rhetoric from objective knowledge might be the earliest possible yet influential starting point in western intellectual history (Eades). Celeste Condit sums up the traditional tension this way: "the 'scientific' view ascribes objective, permanent, and universal status to the facts produced by scientists, whereas the 'sophistic' view supported by many rhetoricians describes facts as products of social conditions, and therefore marked by inter-subjectivity, transience, and situational delimitations" (83). Bridging these epistemological views has, nevertheless, become exigent as scientists increasingly recognize that their work is contingent and dependent on cultural mediation (Nunes; Marcum; Verhoeven et al.) and as rhetorical scholars increasingly call for engagement with the sciences (Brady; Segal) and attend to the multiplicity of other disciplinary calls for interdisciplinary engagement (Shoenberger; Frodeman and Micham).

At this moment, invoking the neurosciences in rhetoric engenders excitement, and it seems that rhetoric follows the larger cultural belief that discoveries about the workings of the human brain retain the power to advance or overturn old perspectives and theories. Jenny Edbauer Rice, for instance, suggests that neuroscientific findings into the non-conscious processes of human affect may advance rhetorical theory. She argues that rhetorical scholars could use interdisciplinary resources from the sciences to consider the way affect transfers back and forth in a room through bodily chemicals and brain signals, and she puts forward the proposition that rhetorical scholars should account for how the body becomes stimulated to enter an affective trajectory that leads to positive or negative dispositions (201). Similarly, John Lynch argues that a logic of representation in rhetoric "undermines a full examination of materiality and the complexity of scientific practice" and that a rhetorical theory invested in materiality must cease the "demonization of scientific materiality as the source of discriminatory attitudes and other detrimental effects" and should turn to the sciences through an analytic lens of articulation to understand how rhetoric acts both out from the body and within the body (435–6).[2]

---

[1] This is not to suggest that sciences of the brain have not been taken up by individual researchers in composition or in English departments more generally. Janet Emig, to take one example, has tried to use neuroscientific insights in the field of composition (see Emig); one might also see Flower and Hayes work on a cognitive process theory of writing, 1981. However, the claim made here is that the trend toward a specific and well-defined engagement in rhetoric as seen in a "neurorhetoric" is a new development.

[2] One assumes that Lynch's call for a new attitude toward the sciences will entail a close inspection and reanimation of the cognitive sciences.

In this chapter, I examine one particularly influential article that follows this approach. In "Identification: Burke and Freud on Who You Are," Diane Davis argues that neuroscience findings can help to undo Kenneth Burke's longstanding and highly influential theory of rhetorical identification, placing identification more in line with Sigmund Freud's non-representationalist "primary identification." To support her argument, Davis draws on neuroscience research into "mirror neurons," a group of neurons that seem to pre-consciously allow "direct access to the mind of others" by enacting a "direct simulation of the observed event" (Gallese et al. 1). In fact, what Davis proposes is a much needed movement in the field of rhetoric away from privileging a rational subject who always makes conscious, logical decisions; her article follows from Joshua Gunn's call to do exactly that. But what proves unique about Davis' work is how she draws in Freud and the contemporary neurosciences to advance this new perspective. Considering how these divergent sources come together to tackle Burke and offer a new view of identification for the field of rhetoric can help illuminate how a neuroscience of rhetoric might proceed.

In this chapter, I examine how Davis deploys neuroscientific work and conceptualizes "identification" by moving between research on mirror neurons and Freud. Davis mediates her affect-based notion of identification through an already familiar humanistic figure—Freud—while also drawing on neuroscience findings about mirror neurons. To make these new claims about the human communication process, Davis had to sidestep consideration of how neuroscience findings themselves depend on rhetorical factors and contingencies.[3]

This is not to suggest that an effort to build new theory from the neurosciences must treat the science as either an inevitably shaky social construction or as an objective absolute truth; neither is it to suggest that attempts to use the neurosciences are ill advised or capable of reversing an intellectual culture where rhetoric becomes once again merely "the title of a doctrine and practice" and no longer the "condition of our existence" (Bender and Wellbery 25). Rather, my examination of Davis' work assumes that both neuroscience and rhetoric are claim makers seeking logical coherence in their respective fields producing knowledge within the confines and affordances of disciplinary practices and logics. Thus, this chapter is positioned at the nexus of a tension in rhetorical scholarship between using neuroscience findings as a motor for new theory and, conversely, acknowledging that those findings are constituted by rhetorical actions. In particular, I examine how rhetorical scholarship uses the symbolic of the neurosciences to pioneer a move beyond the symbolic and account for the motions of the biological—as Burke once called them (Hawhee 158). Diane Davis' use of cognitive neuroscience research on mirror neurons offers

---

[3] An epistemological divide is certainly not the sole reason for the limited amount of direct theorizing from the sciences acknowledged as such; other reasons likely include difficulty in negotiating the complex terminology of the sciences, and difficulty in locating mutual topics of interest that could usefully comment on the key terms from the rhetorical tradition.

some direction for thinking about this issue and working through the tension between a neuroscience of rhetoric and the rhetorical nature of the neurosciences.

The Meaning of Affective Identification

In her article, Davis describes affective identification and argues for its priority in rhetorical scholarship. She begins by reclaiming a Freudian form of identification "always already" happening affectually within the body. She states:

> Freud presents rhetorical studies with another, equally important task: to think the limits of reason by tracking the implications—for society, for politics, for ethics—of a radically generalized rhetoricity that precedes and exceeds symbolic intervention. It seems necessary today, at the very least, to begin exploring the sorts of rhetorical analyses that become possible only when identification is no longer presumed to be compensatory to division. (144–145)

Following from Freud's comment that it is the failure of identification that forms social selves, Davis seeks to make what Burke termed "identification" into an always existent physiological condition that the body itself "fails" to uphold in the conscious "withdrawal of identity" (144). Identification, then, is no longer compensatory to a necessary condition of division between people who are dependent on "a rhetorician to proclaim their unity" (Burke 22); identification is, instead, an underlying function of the human organism.

To support this belief that "Burke censored Freud" (144) and that Freud got it right, Davis turns toward the neurosciences and mirror neurons. Davis situates mirror neuron research as the scientific evidence that supports Freud's pre-rhetorical vision of the mutual, ecological and codependent condition of human existence prior to division. Davis states:

> [W]ho can deny that sense organs and sensory neurons, which operate together not so much at but as threshold, already indicate an excentric structure, an inside-outside similar to a Klein jar or Möbius strip? Mirror neurons, which were discovered in the last decade of Burke's life, offer further confirmation ... This means that the same mirror neurons fire in my brain whether I actually grab a pencil myself or I see you grab one, indicating no capacity to distinguish between my grasping hand and what is typically (and hastily) described as a visual representation of it: your grasping hand ... a mimetic rapport precedes understanding, affection precedes projection. (131)

By situating mirror neuron research as "further confirmation," Davis concludes that the concept of identification should be "rethought" in order to account for this finding: "Identification could not operate among self-enclosed organisms; it would have to belong to the realm of affectable-beings, infinitely open to the other's affection, inspiration, alteration" (133). Such a view of identification does not rely on an a priori distinction between self and other.

From these passages, it is evident that mirror neurons, for Davis, support an ever becoming human organism that develops in tandem with its surrounding environment. Mirror neurons enable Davis to call into question rhetoric as the sole or primary force able to influence as well as constitute subjects at the origins of subjectivity. This desire to think of the human as an organism "always already" identifying with the world is displayed in her choice of terminology. By referring to humans as "affectable-beings," Davis situates what it means to be human as inseparable from the felt effect (the "affect") of a brain teeming with mirror neurons, which comprises the biological mechanism capable of producing, in Davis' view, an embodied identification *before* a self enters the world. Her choice to use the word "affectable-beings" and to avoid the word "human," "person," or "body" reflects the basic ethical potential she sees in a wider recognition of the mutual "exposure" of organisms shaping each other through an entwined process of becoming (*Inessential Solidarity* 6). For Davis, mirror neurons support a view of the human as possessing an unconscious "affectable" physiology evinced by the brain's capacity to "mirror" its environment, a view that works in concert with Freud's view of humans as "affectable-beings" having no a priori division from one another. The ethical charge is to recognize this and to then live it out.

Rhetoric and the Drive to Theorize Affect in the Humanities

What Davis develops is a bold revision of Burke—a well-timed and potent reversal on what has become a doxa of rhetoric. That is, Davis seeks to use Freud and the neurosciences to overturn "all that is considered true, or at least probable" (Amossy 369) by most rhetorical scholars about the processes of human identification. Davis' efforts clearly participate in the ongoing reconsideration of the relationship between rhetoric and science, in line with work by Jenny Edbauer-Rice and John Lynch. But in so doing, she moves toward adopting the belief now circulating in critical cultural theory that "at the heart of 'good theory' lies a prioritization of the biological constitution of being" (Papoulias and Callard 31).

Indeed, the article seems to impel anyone giving weight to symbolic representation in the construction of mental forms to abandon that position and try to account for it through biological processes. Because Davis is interested in a "primary identification" that uproots the "I" from representations and re-roots it in the functioning of an actor's brain, Davis resituates symbolic interaction as dependent on preexisting biological movements. For her, any view of identification as forged out of a discussion of symbols will not offer the information about what identifying with an Other might *most basically* mean (125); for that, a turn to the sciences is needed. Put another way: Davis defers to the neuroscientific lens because that lens does work that a rhetorical one cannot do when it only unpacks the constructedness of things. Davis tends to portray empirical observation as able to transcend its own condition of being constituted within the symbolic, and she holds this position despite the fact that the empirical is only known *through* the negotiation of symbols, as scholars in the rhetoric of science have argued.

Davis' argument reveals a role for the sciences in rhetorical theory. It positions the sciences as existing, on some level, outside of the rhetorician's engagement with symbols. Science, it seems, holds access to a world different from the one available to the rhetorician. Specifically, neuroscience is positioned as being able to more accurately explain human nature since its tools and methodology allow researchers to access the raw material functioning of the brain. To accept this view, neuroscientific interpretations of brain functioning must be imagined, at least provisionally, as unmediated and unfettered by representations so that they can explain more than traditional humanistic interpretations of human behavior. If Davis does not endow a certain privilege to science in this respect, then it is unlikely that she could assert what identification really *is*. For Davis, in other words, the scientific language of the empirical observation receives no explicit treatment as itself rhetorical but is, instead, taken as a supra-rhetorical observation.

Even so, it is likely that Davis recognizes the rhetoricity of the neurosciences. She is, after all, a rhetoric scholar. Yet, in this article, Davis likely chooses not to address it for rhetorical reasons of her own. For one, the article is intended to overturn Burke; thus, Davis needs to assert that the sciences demonstrate a connection with others through pre-conscious neuronal simulation, not through rhetoric. That is the point. And in this case, both types of identification cannot be upheld as "true" since Davis asserts that Burke's notion of identification happening after a built-in division between bodies assumes a homuncular self. Consequently, Burke is understood as too essentialist in his privileging of symbols and needs overturning. But more than this, Davis wants to claim that neuroscience validates Freud's position and can paint the better picture of identification, at least in terms of what identification *is*. In this way, prioritizing the biological over the rhetoricity of neuroscience research serves the rhetorical function of overturning Burke.

Davis' use of neuroscience also participates in a broader conversation within the humanities, mainly a drive to theorize "affect" (see Greg and Seigworth; Hansen; Hawhee; Massumi). Davis' "neuroscience of rhetoric" participates in the ongoing, pervasive attempts to adopt the sciences of the human body and to reposition rhetoric as not always dealing with the rational. Indeed, saying that the non-conscious biological processes of humans are able to make a difference to what human identification can mean for rhetorical scholars reveals the centrality of the body and, accordingly, the sciences of the body, to Davis' interests. Her work, in this way, follows from Jack Selzer's call to embrace materiality, Hawhee's call for rhetorical scholars to take the movement of the body more seriously, and Edbauer-Rice's call to theorize affect.

In short, the argument Davis makes occupies a space now ripe for rhetorical elevation; it carries the "seductive allure" of the neurosciences (Weisberg) over to the issue of identification. In so doing, it makes a popular move toward thinking about the non-symbolic, non-rational processes of the human after a highly publicized and quite impressive decade of cognitive science research pointing to non-conscious influences in everyday life (Damasio; Robinson; Appiah). None of this should, in itself, suggest that Davis is wrongheaded about the issue of identification. The field of rhetoric does need to move beyond what Joshua Gunn described as its underlying "humanist philosophy" wherein rhetors are understood as rational creatures constructing their world through action, and rhetorical scholars should start considering the affective productions of bodies as well as the agency of non-human actors.

Nonetheless, it is worthwhile to consider the intellectual move Davis makes by theorizing affect in rhetoric using neuroscience. Rhetoric scholars would benefit from a more complete recognition of the complexity and controversy surrounding mirror neurons. Further, scholars who draw on neuroscience could better explicate for the audience how exactly their own revisions of rhetorical theory are themselves rhetorical. In Davis' case, it is worth considering why she chooses a limited segment of the neuroscience research to enhance the power and universality of a brain-based affective mechanism. We might also consider further how she employs the word identification in at least two different ways, as both a non-rational, affective condition and a symbolic action.

The following portion of this chapter examines these potential difficulties in Davis' argument, but it also takes the position that her general investment in the non-symbolic motions of the human body is a positive and productive line of inquiry for the field of rhetoric. I ultimately suggest that the terminology she adopts can illuminate a pathway for a more constructive neuroscience of rhetoric in the subfield of neurorhetoric.

Selecting and Deflecting Mirror Neurons

Davis appeals to mirror neuron systems as "further evidence" by which to overturn a Burkean "ontobiological divisiveness." But a more complete overview of the neuroscientific research suggests that mirror neurons may not be as prevalent or important for some people as for others (Dapretto; Oberman et al.), and there remains controversy as to whether or not "mirror neurons" exist in humans at all. Cognitive psychologist Luca Turella and colleagues, for example, carefully detail a long series of cognitive neuroscience studies—including Rizzolatti, Hamzei et al., Grezes et al., Shmuelof and Zohary, among others—arguing that the body of research has not yet shown that mirror neurons exist in humans and that the conclusions are premature and dependent on how one chooses to interpret the results (11–16). Similarly, cognitive neuroscientist Angelika Lingnau and colleagues argue that a series of imitation exercises analyzed in brain imaging studies present no good evidence that mirror neurons exist in humans or enable imitative behavior. In addition, psychologist Jean Decety argues that "several studies claiming to have found MNS [mirror neuron system] activation did not have the appropriate experimental conditions to support such a claim" (206). In a somewhat different register, philosopher Pierre Jacob suggests mirror neurons are not really "mirror neurons" insofar as they do not "mirror" the motor repertoire in another person's brain and are not actually "simulations"; the better explanation, in his view, is that they are predictions of what other people are about to do. Professor of psychological sciences Gergely Cisbra makes a similar argument on similar grounds, contending mirror neurons "do not 'mirror' observed actions with sufficient accuracy for effective simulation" (1). These scientists, then, are themselves using rhetoric to negotiate the grounds for whether mirror neurons exist and what they might mean.

Further, researchers who do accept the existence of mirror mechanisms in humans have pointed out that mirror neurons only fire in special cases, namely when an individual is engaged in specific types of action in a context requiring some specific evolutionary survival response—such as grabbing and eating food—not in every lived context or in reference to every movement (Arbib; Iacoboni et al.). Consequently, any rooting of the self in the Other through this research is somewhat problematic, at least insofar as scholars may claim that human selfhood is dependent on a neuronal "mirroring" of the world and that mirror neuron mechanisms retain some central significance for the material origins of the human self. That said, mirror neurons may still prove radically important, and it is worth recognizing that many of the articles presented in this chapter have appeared since Davis wrote her article in 2008. But the controversy over mirror neurons is, at this point, more important than ever as these neurons are consistently invoked as explanatory mechanisms across numerous fields of study.

In short, although Davis' argument is strongly rooted in Freud's psychological theorizations, the part of her argument that relies on the contemporary neurosciences seems ambitious when compared to the research on mirror neurons and the debate over their typology (Shapiro), function (Arbib; Jacob; Welberg), existence (Cisbra; Lingnau et al.; Turella) and importance for theories of mind (Heyes; Hikcok). Put simply: Davis presents a singular view of mirror neurons and does not call attention to the rhetorical actions of scientists in selling them or arguing for them, nor does she address the possibility of multiple scientific views about what mirror neurons might be or might mean for the analysis of human life. Consequently, it is unclear whether she shifts at the moment of discussing mirror neurons from a constructivist-centered epistemology to a logical positivist one or whether she seeks to make the research compatible with rhetoric by taking the scientific finding as "good enough" of a construction to be incorporated into a rhetorical perspective. Nevertheless, it is clear that the standards for adoption of mirror neuron research, or any neuroscience research for that matter, would benefit from a rigorous and extended engagement with the literature.

Doubling Identification

A limited deployment of neuroscientific research on mirror neurons can lead to a (perhaps unintentional) "neuroessentialist" version of the self that moves close to equating "subjectivity and personal identity to the brain" (Racine et al. 160). In privileging the mirror mechanism, what drops out is a discussion of the conscious, lived human experience, which remained so central to Burke's version of identification. To show how there are two different identifications being compared and discussed in Davis' article, it should be noted that Davis first argues that an acceptance of Burkean identification requires scholars "also to presume—as the condition for identification—a subject or ego who knows itself as and through its representations" (127). And this is precisely why Burke's theory of identification becomes untenable for Davis—it does not reject ontological foundations and essential humanisms. However, this may not be the case.

For Davis to effectively argue that the self is not divided *first* but is, rather, a dynamic product of the environment wrapped up in an always already identification, she appeals to both Freud and to the contemporary neurosciences. To do so, though, she ends up suspending phenomenal experiences of the body and conscious experience and upholding a body–brain dichotomy. In this way, the brain's automatic processes of environmental simulation become the sole factor in thinking about what the self *is* at its most basic. In short, by prioritizing biological processes, Davis shifts Burke's sense of identification away from a thoroughly embodied and conscious engagement with rhetoric. She, rather, seeks to remake "identification" into the neurological coding of visual actions onto the motor cortex, a process described by Leonardo Fogassi and Vittorio Gallese—one of the researchers who discovered mirror neurons—as "in no way obvious" and occurring below the level of consciousness (*Mirror Neurons and the Evolution of Brain and Language* 13).

Of course, Davis' goal is to assert that identification is most primarily a non-rational, pre-conscious experience, so saying that she overlooks Burke's focus on encounters with the symbolic in an active social sphere might sound like an odd critique. However, Davis builds her argument on the assertion that Burke likely engaged in a psychological suppression of Freud's insights (141–142) even though it seems more likely that Burke (characteristically) intended to bridge Freudian concepts with a more sociological view of lived experience. In *The Rhetoric of Motives*, Burke speaks about bodies as being *already* physiologically divided from each other before the actions of they engage in. For Burke, bodies form a division that stand as the beginning of a separated selfhood that later becomes conscious and immediately embedded in the reflective position of asking "who am I?" And that question seems to be his prime concern when discussing the self.

Burke's position on the primacy of a division of bodies becomes evident when the reader recognizes that a discussion of war, Hitler, and rival factions surround his famous passage on identification (22–23). The emphasis is on conscious, lived experience and on active self-formation in awareness; the body before that time may not be inhabited by a homuncular manifestation. From all indications, the Burkean conception of the Self *before* representation is likely nothing more than an entity with "uniqueness… in itself and by itself" (21). The phrasing Burke uses here could just as likely point to the body's capacity to become uniquely itself and does not necessarily indicate a presumption about a preeminent, homuncular Self. In fact, it is telling that Burke maintains a clear interest in the human body and its non-symbolic affective motions (Hawhee, 51–54, 83–85) but still considers "identification" to be a cultural effect independent of any claim about those affective motions. This is telling not because it reveals, as Davis assumes, his essentialist ontological position but because it reveals what he means when he says "identification."

Burke's strong emphasis on identification as intimate with representation can be understood by considering his discussion of John Milton's poem about Samson, who "slew himself in slaying enemies of the Lord" (19). Starting this way in his section on identification suggests that killers are "'essentially identifiers'" (20). Further, Burke suggests that Samson's destruction constitutes self-destruction insofar as his identity as "killer" depends on his relationships to those whom he kills. Despite this inherent recognition of an entwined identity, Burke seems to suggest that the presencing of others enables one's own condition, and it is, thus, through presencing or reference that a self is forged. Consequently, it is not surprising that immediately after this passage on Samson, Burke says, "A is not identical with his colleague, B," asserting that A must be persuaded to see their interests as joined (20). Overall, then, it is clear that Burke thinks from the conscious condition of social bodies for theorizing identification, even as he recognizes the need for an Other to prop up an identifying self. Therefore, it appears likely that non-conscious processes turning below a level of awareness simply would not persuade Burke that affective experience is a properly constitutive form of "identification" not subsumed by the conscious experience of the division of bodies as skin-bound organisms appearing before each other.

In brief, Davis' "identification" retains little similarity to Burke's rhetorical "identification" because of the way they both set out to define the term. And Davis' argument does not necessarily undercut Burke's position on identification since her view does not show that a conscious Burkean self-formation after symbolic experience is less important to selfhood, at least as Burke defined it in terms of lived experience, than what happens in an affective simulation in the brain. Her argument would need to show how affect wraps over or within Burke's concern with lived experience such that conscious experience of a skin-bound body is already an affectual identification that makes division and symbolic play from that conscious position less important to the processes of identifying. In addition, she would need to directly address why the origins of a self through an egoist disidentification would not support Burke's position over her own.

In short, Davis seems to say something more or something in a different register than Burke. And seeing the difference exposes how terms like "identification" shift with their context of use or, as Bruno Latour might say, reassemble with the enrollment of new actors and alliances that assemble terms. In this case, Davis "assembles" identification by integrating Freud and current discussions of affect and neuroscience. In contrast, Burke "assembled" identification as a conscious division. For Burke, the affective mechanism can never again precede the conscious experience; consciousness disables the "always already-ness" of affect. In short, the two meanings of identification (Davis' and Burke's) seem more complimentary than competitive.

Embracing the Neurosciences in the Spirit of the Many

The reason for Davis's intense emphasis on affective identification lies in her effort to bring together cultural theorists, neuroscientists, and psychologists around the notion that "the singular being is not enclosed in a form and cannot appear or even exist alone" but is "by definition shared" and that always exposes itself as "an inappropriable outside that constitutes it, affects it, alters, prior to and in excess of symbolic intervention" (2010, 6–8). Her efforts are laudable in this regard, but her arguments might also be clarified through further engagement with the neuroscientific research and further discussion of the weight that Burke seems to hand to conscious experience over and above the automatic motions of brains. In short, the rhetoric of neuroscience could inform Davis' understanding of identification.

By the same token, a neuroscience of rhetoric could learn from Davis' drive to expand the perspective of the human. That is, a rhetoric of the many and not of the few, one that exposes the way it forms itself in making choices to select and deflect, one that stands open to a complete, even totalizing revision, will not only be the most rhetorical, but the most persuasive. The move to see scientific work as an equally legitimate form of knowledge making does not require tiptoeing stealthily between a constructivist epistemology and a logical positivist one; bridging an epistemological gap does not require deference to the sciences or a rejection of rhetoric. In fact, if rhetorical theory is to fold in work from the sciences, then it need not see itself as superior to science's claims to discovery but be open to the possibility that scientific findings might enrich or undo rhetorical theories. This is not a rejection of rhetoric's epistemology but an embrace of it in a way that invites scientific work to participate in the coherence of rhetoric.

If Davis' article is any indication, then a turn to affect in rhetoric may uproot liberal humanism's monochromatic assertion that life moves freely and independently of the surrounding environment. Yet, this inquiry into affect must be fully engaged with the scientific research it invokes, which may include forging partnerships with scientific researchers. An alternate strain of scientific research offers another direction for reconceptualizing being-in-the-world. Researchers such as Damasio argue that conscious experience is equally non-rational, and the turn to affect will surely find productive cooperation with an expanded and more complex view of consciousness as always inherently emotional, limited in its attentive capacities, and in need of different kinds of coherence. Davis takes a much needed step in rhetoric by exploring the non-rational. Finding ways to compare what we think we knew to what we think we now know in rhetoric about the conditions and formations of life is, first, a matter of someone like Davis stepping into new territory, and then, second, a matter of turning over the ground.

Exposing Ourselves

In their article examining how humanities disciplines take up neuroscience research, Jenell Johnson and Michelle Littlefield argue that many disciplines (such as sociology and philosophy) treat the neurosciences as big-S Science itself and not as a series of practices and interpretations. Davis' article can be viewed as one potential set of rhetorical actions employed when scholars turn to the neurosciences as evidence for a new theoretical position. It may not be unexpected to see articles with the following rhetorical tendencies:

- neuroscience research is positioned as universal, objective or non-rhetorical
- neuroscience research is made to appear stable, and disagreements in the neuroscience community are not readily acknowledged
- neuroscience research is positioned as the spokesperson for nature and operates as an explanatory mechanism for questions about human activity over and above humanistic or social scientific interpretations
- neuroscience research is compared or situated in reference to a source more familiar to the audience (such as Sigmund Freud or, perhaps, Daniel Stern or Maurice Merleau Ponty, whomever the situation may conjure.) Or the neuroscience research is used to prove that theorist was right and that neuroscience can legitimize old and forgotten claims
- theoretical terms become vague or take on multiple meanings when the neuroscience is used to revise the meaning of an existing term in the field.

These findings serve as a starting point for understanding the ways neuroscience research is usefully incorporated into non-neuroscience fields and shaped to develop different and sometimes better or timely answers to field-specific questions. Ultimately, these findings prove suggestive toward interdisciplinary theory making involving the neurosciences and may point to an existing tendency to overturn humanistic and social scientific theories when in conflict with the interpretations of neuroscientific data—despite the instability of the neuroscience in question, the multiple interpretations applied, and the possibility of mutual coexistence between new and old theories.

The larger question, then, is what is the state of theoretical development in the humanities and social sciences? Does interdisciplinary research and theory making require deference to neurobiology? I would tend to reject the either–or dichotomy and to say not necessarily and to add that deference might be required at times. But I would also argue that a turn to the neurosciences requires, as Davis says, the embrace of "a task given to 'us,' in the name of solidarity, to expose exposedness" and to make the many appear in the mutual production of logical coherence (2010,

8). This is not a Derridean call to perpetual deconstruction but an expression of self-nudity needed from all analytic positions at a time when it is tempting to wear some of the emperor's new clothes.

But calling the neurosciences our contemporary naked emperor parading with faux confidence before serfs and servants in other disciplines requires looking down and seeing one's own nakedness, the nakedness we share, of sense making dependent on what seems correct at a time and place from the particular foci of history and the material orientations we have to a complex world. Thus, it is important to recognize that the contemporary enthusiasm for analytics stressing entwined interconnectedness is inseparable from and imbued with a networked logic at a time and place when more and more things are physically connected through new digital infrastructures. This inherently political drive within the affective analytic—to amplify the many-together at a foundational level louder than the one-alone in a context of growing global relations—may not dismiss the dependence of meaning making on reference and on division amongst "bodies" as much as it recognizes the need, now more than ever before, for global peace and unity in this radically interdependent technological world.

In so doing, the affective distances itself from a postmodern turn that has for so long lingered in the humanities like a shadowy abyss at the center of thought about human life. The focus on the tangible raw meat of the material body, in rejection of Burke, must stand no longer as the foundational division in human life but as the fundamental unity bridging the abyss of an ever shifting postmodern scaffold of difference; the biological material that seems to exist without our representations (but has, nevertheless, endowed us with them as modes of shaping this world) is itself represented as the escape hatch to tunnel beneath our descriptions of conscious experience so that we can get at something indivisible. The problem, it seems, then, is the mind–body—that is to say a historical understanding of what a human *is*. But if the brain–body is enrolled as the best solution to this historical (mind–body) problem, then what a poor solution the brain is. In other words, the appeal points to the materiality of the brain to resolve a problem that may not be located or perpetuated there. To complicate the matter, the appeal itself requires representations while claiming to be independent of representation. As the case of the mirror neurons attests—with myriad interpretations and methodological disagreements— the study of the brain can only lead back to division again and again. And regardless of the fact that there is a material substance to the brain that exerts force and needs to be accounted for, any description of it cannot very easily claim to be a description of it before the description of it. As a result, claims about preexistence, as opposed to claims about simultaneity and mutual feedback loops, call attention to their own construction in ways damaging to the point of the claim.

Turning to the neurosciences may be productive for many pursuits, but it may simply be the wrong place to look for evidence of a unity that preexists symbolic intervention.

References

Amossy, Ruth. "Introduction to the Study of Doxa." *Poetics Today* 23.3 (2002): 369–394. Print.

Appiah, Kwame Anthony. *Experiments in Ethics*. Cambridge: Harvard University Press, 2010. Print.

Bender, John, and David E. Wellbery. *The Ends of Rhetoric: History, Theory Practice*. Stanford: Stanford University Press, 1990. Print.

Brady, Ann, Robert R. Johnson, and Charles Wallace. "The Intersecting Futures of Technical Communication and Software Engineering: Forging an Alliance of Interdisciplinary Work." *Technical Communication* 53.3 (2006): 317–325. Print.

Buccino, Giovanni, et al. "Action Observation Activates Premotor and Parietal Areas in a Somatotopic Manner: An fMRI Study." *European Journal of Neuroscience* 13 (2001): 400–404. Print.

Burke, Kenneth. *A Rhetoric of Motives*. New York: Prentice Hall, 1950. Print.

Ceccarelli, Leah. *Shaping Science with Rhetoric: the Cases of Dobzhansky, Schrödinger, and Wilson*. Chicago: University of Chicago, 2001. Print.

Cisbra, Gergely. "Mirror Neurons and Action Observation. Is Simulation Involved?" *What Do Mirror Neurons Mean? Interdisciplines Conference*, 2004. Web. 22 October 2011. <http://www.interdisciplines.org/medias/confs/archives/archive_8.pdf>.

Condit, Celeste. "How Bad Science Stays That Way: Brain Sex, Demarcation, and the Status of Truth in the Rhetoric of Science." *Rhetoric Society Quarterly* 26.4 (1996): 83–109. Print.

Damasio, Antonio. *The Feeling of What Happens Body and Emotion in the Making of Consciousness*. New York: Harvest, 1999. Print.

Damasio, Antonio. *Self Comes to Mind*. New York: Pantheon, 2010. Print.

Dapretto, Mirella, et al. "Understanding Emotions in Others: Mirror Neuron Disfunction in Children with Autism Spectrum Disorders." *Nature Neuroscience* 9.1 (2006): 28–30. Print.

Davis, Diane. "Identification: Burke and Freud on Who You Are." *Rhetoric Society Quarterly* 38.2 (2008): 123–147. Print.

——. *Inessential Solidarity: Rhetoric and Foreigner Relations*. Pittsburgh: University of Pittsburgh Press, 2010. Print.

Decety, Jean. "To What Extent is the Experience of Empathy Mediated by Shared Neural Circuits?" *Emotion Review* 2.3 (2010): 204–207. Print.

Eades, Trent. "Plato, Rhetoric, and Silence" *Philosophy and Rhetoric* 29.3 (1996): 244–258. Print.

Emig, Janet. "Writing as a Mode of Learning." *College Composition and Communication* 28.2 (1977): 122–128. Print.

Fahnestock, Jeanne. "Rhetoric in the Age of Cognitive Science." *The Viability of Rhetoric*. Ed. Richard Graff. New York: State University of New York Press, 2005. 159–179. Print.

Flower, Linda, and John R. Hayes. "A Cognitive Process Theory of Writing." *College Composition and Communication* 32:4 (1981): 365–387. Print.

Fogassi, Leonardo, and Vittorio Gallese. "The Neural Correlates of Action Understanding in Non-Human Primates." *Mirror Neurons and the Evolution of Brain and Language*. Amsterdam: John Benjamins, 2002. 13–35. Print.

Frodeman, Robert, and Carl Mitcham. "New Directions in Interdisciplinarity: Broad, Deep, and Critical." *Bulletin of Science, Technology and Society* 27.6 (2007): 506–514. Print.

Gallese, Vitorrio, Christian Keysers, and Giacomo Rizzolatti. "A Unifying View of Social Cognition." *TRENDS in Cognitive Sciences* 8.9 (2006): 396–403. Print.

Gibbons, Michelle. "Seeing the Mind in the Matter: Functional Brain Imaging as Framed Visual Argument." *Argumentation and Advocacy* 43 (2007): 175–188. Print.

Gregg, Melissa, and Gregory J. Seigworth. *The Affect Theory Reader*. Durham: Duke University Press, 2010. Print.

Grezes, Julie, et al. "Activations Related to 'Mirror' and 'Canonical' Neurons in the Human Brain: An fMRI Study." *Neuroimage* 18 (2003): 928–937. Print.

Gross, Alan G., and William M. Keith. *Rhetorical Hermeneutics: Invention and Interpretation in the Age of Science.* Albany: State University of New York, 1997. Print.

Gruber, David, et al. "Neuroscience and Rhetoric: Engagement and Exploration" *POROI* 7.1 (2011): 1–12. Print.

Gunn, Joshua. "Review Essay: Mourning Humanism, or, the Idiom of Haunting." *Quarterly Journal of Speech* 92.1 (2006): 77–102. Print.

Hamzei, Farsin, et al. "The Human Action Recognition System and its Relationship to Broca's Area: An fMRI study." *Neuroimage* 19 (2003): 637–644. Print.

Hawhee, Debra. *Moving Bodies: Kenneth Burke at the Edges of Language.* Columbia, SC: University of South Carolina, 2009. Print.

Heyes, Cecilia. "Mesmerizing Mirror Neurons." *NeuroImage* 51 (2010): 789–791. Print.

Hickok, Gregory. "Eight Problems with Mirror Neuron Theory as Action Understanding in Monkeys and Humans." *Journal of Cognitive Neuroscience* 21.7 (2008): 1229–1243. Print.

Iacoboni, Marco, and Marella Dapretto. "The Mirror Neuron System and the Consequences of its Dysfunction." *Nature Reviews Neuroscience* 7 (2006): 942–951. Print.

Jack, Jordynn. "What Are Neurorhetorics?" *Rhetoric Society Quarterly* 40.5 (2010): 406–411. Print.

Jack, Jordynn, and L. Gregory Applebaum. "'This Is Your Brain on Rhetoric': Research Directions for Neurorhetorics." *Rhetoric Society Quarterly* 40.5 (2010): 412–439. Print.

Jacob, Pierre. "What do Mirror Neurons contribute to Human Social Cognition?" *Mind and Language* 23.2 (2008): 190–223.

Johnson, Davi. *Brain Science: Neuroscience and Popular Media.* Rutgers University Press, 2011. Print.

Johnson, Jenell, and Melissa Littlefield. "Lost and Found in Translation: Popular Neuroscience in the Emerging Neurodisciplines." *Advances in Medical Sociology.* Eds. Martyn Pickersgill, & Ira Van Keulen. London: Emerald Insight. In Press.

Keränen, Lisa. *Scientific Characters: Rhetoric, Politics, and Trust in Breast Cancer Research.* Tuscaloosa: University of Alabama, 2010. Print.

Latour, Bruno. *Reassembling the Social.* Oxford: Oxford University Press, 2005. Print.

Lingnau, Angelika, Benno Gesierich, and Alfonso Caramazza. "Assymetric fMRI Adaptation Reveals No Evidence for Mirror Neurons in Humans." *Proceedings of the National Academy of Sciences.* 106.24 (2009): 9925–9930. Print.

Lynch, John. "Articulating Scientific Practice: Understanding Dean Hamer's 'Gay Gene' Study as Overlapping Material, Social and Rhetorical Registers." *Quarterly Journal of Speech* 95.4 (2009): 435–465. Print.

Marcum, James A. "Instituting Science: Construction of Scientific Knowledge?" *International Studies in the Philosophy of Science* 22.2 (2008): 185–210. Print.

Massumi, Brian. *Parables for the Virtual: Movement, Affect, Sensation.* Durham, NC: Duke University Press, 2002. Print.

Maturana, Humberto. "Science and Daily Life: The Ontology of Scientific Explanations." *Selforganization: Portrait of a Scientific Revolution.* Eds. W. Krohn, Gunter Küppers, and H. Nowotny. Dordrecht: Reidel Publishing (1980): 12–35. Print.

Merleau-Ponty, Maurice. *Phenomenology of Perception.* New York: Humanities, 1962. Print.

Miller, Carolyn. "Kairos in the Rhetoric of Science." *A Rhetoric of Doing: Essays Written in the Honor of James L. Kinneavy.* Eds. Stephen P. White, Neil Nakadate, & Roger D. Cherry. Carbondale: University of Illinois Press, 1992, 210–37. Print.

Mol, Anne Marie, and John Law. "Regions, Networks and Fluids: Anaemia and Social Topology." *Social Studies of Science* 24.4 (1994): 641–671. Print.

Nunes, Terezinha. "The Sociocultural Construction of Implicit Knowledge." *Cognitive Development* 18.4 (2003): 451–454. Print.

Oberman, Lindsey M., et al. "EEG Evidence for Mirror Neuron Dysfunction in Autism Specturm Disorders." *Cognitive Brain Research* 24.2 (2005): 190–198. Print.

Papoulias, C., and F. Callard. "Biology's Gift: Interrogating the Turn to Affect." *Body and Society* 16.1 (2010): 29–56. Print.

Prelli, Lawrence. *A Rhetoric of Science: Inventing Scientific Discourse*. Columbia: University of South Carolina Press, 1989. Print.

Racine, Eric, Ofek Bar-Ilan, and Judy Illes. "Science and Society: FMRI in the Public Eye." *Nature Reviews Neuroscience* 6.2 (2005): 159–164. Print.

Rice, Jenny Edbauer. "The New 'New': Making a Case for Critical Affect Studies." *Quarterly Journal of Speech* 94.2 (2008): 200–212. Print.

Rizzolatti, Giacomo, et al. "Localization of Grasp Representation in Humans by PET: 1. Observation Versus Execution." *Experimental Brain Research* 111 (1996): 246–252. Print.

Robinson, Jenefer. *Deeper Than Reason*. Oxford: Clarendon Press, 2005. Print.

Segal, Judy. "Interdisciplinarity and Bibliography in Rhetoric of Health and Medicine." *Technical Communication Quarterly* 14.3 (2005): 311. Print.

Selzer, Jack. "Habeas Corpus: An Introduction." *Rhetorical Bodies*. Madison: University of Wisconsin Press, 1999. 3–15. Print.

Shapiro, Lawrence. "Making Sense of Mirror Neurons." *Synthese* 167 (2009): 439–456. Print.

Shmuelof, Lior, and Ehud Zohary. (2006). "A Mirror Representation of others' Actions in the Human Anterior Parietal Cortex." *Journal of Neuroscience* 26, 9736–9742. Print.

Schoenberger, Erica. "Interdisciplinarity and Social Power." *Progress in Human Geography* 25.3 (2001): 365–382. Print.

Slaby, Jan. "Steps Towards a Critical Neuroscience." *Phenomenology and Cognitive Sciences* 9 (2010): 397–416. Print.

Stamenov, Maksim, and Vittorio Gallese. *Mirror Neurons and the Evolution of the Brain*. Amsterdam: John Benjamin Publishing, 2002. Print.

Turella, Luca, et al. "Mirror Neurons in Humans: Consisting or Confounding Evidence?" *Brain and Language* 108 (2009): 10–21. Print.

Umiltà, M. A., et al. "'I Know what You are Doing': A Neurophsyiological Study." *Neuron* 32 (2001): 91–101. Print.

Verhoeven, Ludo, Wolfgang Schnotz, and Fred Pass. "Cognitive Load in Interactive Knowledge Construction." *Learning and Instruction* 19.5 (2009): 369–375. Print.

Weisberg, Deena Skolnick, et al. "The Seductive Allure of Neuroscience Explanations." *Journal of Cognitive Neuroscience* 20.3 (2008): 470–477. Print.

Welberg, Leonie. "Mirror Neurons: Toward a Clearer Image." *Nature Reviews Neuroscience* 9 (2008): 888–889. Print.

Toward a Rhetoric of Cognition

Daniel M. Gross

In their introduction to this volume, Jordynn Jack and L. Gregory Appelbaum call for collaborative, interdisciplinary research that avoids the pitfalls of neurorealism, neuroessentialism, and neuropolicy, all of which "tend toward uncritical fetishization of the brain as a scientific object divorced from its historical and rhetorical context" (409). Likewise in her recent *Nature* review of Simon Baron-Cohen's 2011 book on empathy, Michigan professor of cognition and cognitive neuroscience Stephanie Preston cautions readers against making too much of Baron-Cohen's "empathy circuit" and its brain localization, while instead promoting an interdisciplinary framework that combines neuroscientific knowledge with findings from social and political science designed to better capture the "richness of the human context" (416).

For Preston, "context" refers to the disciplinary expertise of social and political scientists, but, in what follows, I will argue for the relevance of disciplinary expertise emerging from the humanities and rhetorical studies not only in the interpretation of neuroscience findings, but in experimental design. Interdisciplinary research, Jack and Appelbaum suggest, is already emerging around two approaches they identify as:

1 the *rhetoric of neuroscience*, which appears as a familiar subcategory of the rhetoric of science insofar as it examines the "modes, effects, and implications of scientific discourse" such as the 2010 effort to brain-image "yes" and "no" (headline: "Vegetative patient 'talks' using brain waves")
2 the *neuroscience of rhetoric*, which uses neuroscientific research to reconsider familiar rhetorical issues in language, persuasion, and communication, such as the relationship between pathos and logos, or to explore new territory such as physiological questions about how brain differences might influence communication (412; see also Gruber, this volume).

Both approaches focus on discourse but I think there is at least one more approach worth identifying.

This response will introduce a third, and, I believe, distinct, approach to collaborative, interdisciplinary research shared by neuroscientists and rhetoricians, which I will call the *rhetoric of cognition* to deemphasize the nervous system, and to highlight the embeddedness of cognition as perceiving, conceptualizing, knowing-that, and knowing-how: i.e, modalities of being-in-the-world that can significantly exceed discourse. This third approach considers seriously the brain as a scientific object not divorced from, but instead embedded in the "historical and rhetorical context" mentioned by Jack and Appelbaum, although this embeddedness is addressed at the level of experimental design as well as data interpretation. I am currently pursuing this type of work with empathy specialist Preston—work that finds some sympathy in the emerging field of "situated cognition" studies, which already has rhetorical implications insofar as it foregrounds serious consideration of context or what is more frequently referred to as the "situation." But what situation exactly? Characterized how? And in what relevant ways related to or different from the rhetorical situation as it is treated in our canonic literature on the topic? With what implications for a rhetoric of cognition?

In what follows I will briefly address these questions by examining an influential study (Phelps et al.) that correlates unconscious racial prejudice with amygdala response. Along the way I will also indicate openings for rhetoric to play a role in situated cognition studies; drawing on essays by Paul Griffiths and Andrea Scarantino, and Christy Wilson-Mendenhall et al. I will suggest how a more thorough engagement with rhetoric will improve studies in situated cognition.

In a widely cited article that appeared in the *Journal of Cognitive Neuroscience*, New York University cognitive scientist Elizabeth A. Phelps and her colleagues used functional magnetic resonance imaging (fMRI) to examine the neural response of "White American" subjects who viewed images of African Americans and "White Americans." Phelps et al. describe their study in social cognition as being motivated by the remarkable difference between, on the one hand, a steady decline in self-reported racial prejudice, and, on the other hand, persistent "unconscious" racial prejudice directed at Black people (729). One of two primary goals in this study was to examine the neural correlates of responses to racial groups, focusing on the amygdala because it appears to be involved in emotional learning especially in relation to fear, memory, and evaluation (729). Using fMRI brain-imaging technology, the Phelps group investigated correlations among scores on the Modern Racism Scale[1] which is commonly used to measure conscious, self-reported beliefs and attitudes toward Black Americans (736), unconscious measures of

---

[1] They asked for agreement or disagreement with items like the following, with lower scores representing pro-Black and higher scores representing anti-Black beliefs and attitudes. "Discrimination against blacks is no longer a problem in the United States"; "It is easy to understand the anger of black people in America" (736).

startle response and bias (730),[2] and fMRI-derived amygdala activity in White American subjects responding to Black and White male faces with "neutral" expressions. In Experiment 1, the faces presented belonged to individuals who were unfamiliar to the subjects. In Experiment 2, the faces belonged to "famous and positively regarded" Black and White individuals (730) including Arsenio Hall, Bill Cosby, Magic Johnson, Martin Luther King, Jr., Colin Powell, and Conan O'Brien, Tom Cruise, Larry Bird, John F. Kennedy, Norman Schwarzkopf (736). According to Phelps, these two studies showed for the first time that members of Black and White social groups can "evoke differential amygdala activity" and that "this activity is related to unconscious social evaluation" (734). In Experiment 1, the strength of amygdala activation to Black versus White faces was correlated with two unconscious measures of race evaluation, but not with conscious expression of race attitudes. In Experiment 2, these patterns were not produced when the faces observed belonged to familiar and positively regarded Black and White individuals (734).

So where exactly might rhetoricians engage with this study, and what are the implications for a rhetoric of cognition? Phelps et al. design the study around the discrepancy between verbalized race attitudes and unconscious race attitudes manifested by way of indirect word/image associations and startle response. Indeed, the researchers experimentally distinguish the "racial bias" they measure from the purposeful and conscious action of discrimination verbally attested by particular subjects (730), instead attributing bias to patterns of association built into a particular culture (and instantiated by way of the subject).

By drawing on critical race theory and rhetorical studies, we might say that the White, avowedly pro-Black subject who startles more quickly, and negatively associates more readily when presented with an unfamiliar Black face, is, in part, *subjectified* by racist culture that sometimes shows up in brain physiology but cannot be located exclusively in the brain. Instead, racist culture is located primarily in the racist institutions that exceed any particular subject and can persist without any particular kind of incorporation. In this case, both sides of the equation are necessary for the study to matter: without the physiology, claims of declining racism might be overstated due to the ideological prestige of antiracist discourse, and without rhetorical studies broadly conceived, such physiological evidence could never show up in the first place because we would not know what we were looking for. After all, "race" doesn't originate in the brain but rather in a social world where natural kinds are less important than structures of power and persuasion: i.e., structures of rhetoric (see Jackson, this volume). Among other things, the Phelps study demonstrates how racist institutions "persuade" subjects,

---

[2] 1) eyeblink startle response comparisons and 2) Implicit Association Test (IAT) where subjects categorize "Black" and "White" faces while simultaneously categorizing "good" words (joy, love, peace) and "bad" words (cancer, bomb, devil).

i.e., constitute subjects as racist and racialized bodies, precisely under the cover of discourse that would deflect the very appearance of racialized subjectivity.

Rhetoric scholars might also be interested in how mediation shows up in Experiment 2, where faces belonging to "familiar and positively regarded" Black and White individuals significantly mitigate the racism identified in Experiment 1. But instead of emphasizing like Phelps et al. the antiracist implications of this finding, which would suggest that familiarity mitigates racism, a rhetorician might underscore the heavily mediated quality of each experimental stage. This would include category selection ("Black" and "White" that are sociohistorical categories not natural kinds), image selection ("Black" and "White" image attribution that will vary dramatically according to the cultural situation of the laboratory and its experiment), and also "familiarity," which in this case has nothing to do with face-to-face encounters, personal acquaintance, or family or communal relationships, and everything to do with mediation as *media* that locates and dis-locate images according to complex regimes of publicity and capital of the sort studied thoroughly by media and rhetoric scholars (see Johnson, this volume). Hence one might question, for instance, the Phelps presentation of stimuli that equates a list of Black and White individuals whose faces are supposed to portray rough equivalents in degrees of fame, age, and achievement (733). For instance, only a broader perspective can foreground the fact that, although Martin Luther King, Jr. and John F. Kennedy might provide facial images roughly equivalent within strict parameters defined by the experiment, fame, achievement and even longevity are unevenly distributed across Black and White populations, which means at least that a famous and familiar Black face carried with it into the laboratory certain kinds of markedness, whereas the supposedly equivalent White face did not. To put this another way, Phelps may be correlating *celebrity* with unconscious racial attitudes, not familiarity, which is certainly interesting in terms of media and rhetorical studies, but does not exactly speak to the Phelps' objective as stated. Moreover, such differentiated mediation suggests, for the sake of my argument, that the cognition study by Phelps must be "situated" with greater rhetorical specificity if it is to produce results that speak to causality.

Indeed Phelps et al. readily admit that, although they identify significant correlations between amygdala activity and indirect behavioral measures of racial bias, "these data cannot speak to the issues of causality" (734), which is not to say they do not interpret their data in a manner that has interesting social consequences:

> Our own interpretation is that both amygdala activation as well as behavioral responses of race bias are reflections of social learning within a specific culture at a particular moment in the *history of relations between social groups* … Unless one is socially isolated, it is not possible to avoid acquiring evaluations of social groups, just as it is not possible to avoid learning other types of general world knowledge. Having acquired such knowledge, however, does not require its conscious endorsement. Yet such evaluations can affect behavior in subtle and often unintentional ways. (Phelps et al.)

A rhetorical perspective might focus further attention on the "history of relations" that becomes indispensable for the interpretation of certain kinds of amygdala response. Also note the follow-up sentence about how we are implicated in those histories of relation that compose part of what Phelps et al. call "general world knowledge," where a rhetoric scholar might highlight knowledge that exceeds the sum total of propositional content to include ways of being (knowhow *à la* Heidegger). A rhetorical perspective might also consider knowledge that is not housed in the brain per se, but instead in environmental affordances (à la Gibson) such as structures of racial inequity including explicit law and implicit *habitus*, or the media that once offered Arsenio Hall as a particular kind of friendly face bridging the divide between young Black people and others. After all, we can presume that the experimental subjects had never met Arsenio Hall face to face, so whatever that encounter in the laboratory produced in terms of comparative data, mediation has to be considered at both ends of the experiment including design and interpretation.

Particularly promisingly for rhetoricians, Phelps' study tinkers with social context by changing around the familiarity variables or the goal orientations, to demonstrate the context dependence, or what Paul Griffiths and Andrea Scarantino call an "affordance" profile, of certain kinds of amygdala response. Instead of working in the laboratory, Griffiths and Scarantino track "emotions in the wild" in the effort to shift theoretical focus to neglected phenomena such as anger in a marital quarrel or embarrassment while delivering a song to an audience (438). Emotional content, they argue, has a "fundamentally pragmatic dimension, in the sense that environment is represented in terms of what it affords the emoter in the way of skillful engagement with it" (441). In an effort to theorize cognition beyond the brain, Griffiths and Scarantino consider the "active contribution of the environment" such as confessionals in churches that enable certain kinds of emotional performance that they call synchronic scaffolding, and the broader Catholic culture that supports the development of the ability to engage in the emotional engagements of confession, which they call diachronic scaffolding (443). As I do in *This Secret History of Emotion*, they also consider material factors including "emotional capital" such as the emotional resources associated with having a specific social status, gender, disability, etc. (444; see also Pryal, this volume).[3] Phelps et al. take up context dependence by comparing familiar to unfamiliar faces, but certainly other kinds of contextual factors could be introduced in such a study.

Rhetoric scholars might also pay attention to the specific brain regions invoked in studies of cognition. Notably, Phelps qualifies the locational specificity of her

---

[3] "Affordance" theory derives most importantly from the work of Gibson and biologist Jakob Johann von Uexküll, whose 1909/1924 *Umwelt und Innenwelt der Tiere* had an influence on Martin Heidegger's 1927 *Sein und Zeit*. On emotional "scaffolding," see, for instance, Vygotsky.

studies. In one instance, she explains that a similar response profile is produced by someone with amygdala damage, which suggests that the response she is studying is very complex and implicates factors that extend beyond a particular region in the brain hardwired for a narrow range of experiences such as "fearing outsiders" or snakes or some such. How complex is precisely the territory of neuroscientists, and how far beyond is the territory of rhetoricians broadly conceived.

While their work clearly involves contextual and rhetorical factors, Phelps and her colleagues seem to recognize these elements in a rather limited sense. In a related 2008 article, Stanley and Phelps et al. talk about fusing disciplines: "The fusion of neurobiological and psychological findings into a single model of how implicit attitudes are represented, express, and regulated is critical for understanding how these attitudes affect, and are affected by, our social interactions and environment."[4] They gesture only weakly toward social interactions and environment, the "context of a situation," and at this point Wilson-Mendenhall and Barsalou's work on situated cognition can help focus the discussion.

In a summative essay "Grounding Emotion in Situated Conceptualization" (2011), C. D. Wilson-Mendenhall et al. critique the "basic emotion" approach made famous by Paul Ekman, where the central hypotheses are that "emotions reflect inborn instinct, and that the mere presence of relevant external conditions triggers evolved brain mechanisms in in a stereotyped and obligatory way (e.g., a snake triggers the fear circuit)" (1105). Interestingly for rhetoricians as an example of what Jack and Appelbaum call the *neuroscience of rhetoric*, a key component of the Wilson-Mendenhall and Barsalou's argument postulates that "situated conceptualization" should include abstract concepts like *convince* and emotional concepts like *fear* and *anger*. "Abstract concept such as *convince*," they explain: "typically refer to an entire situation, not just to part of one, such that an entire situated conceptualization represents them. *Convince*, for example, integrates an agent, other people, an idea, communicative acts, and possible changes in belief, all organized with a variety of relations, such as the relation of one person having an idea, talking with another, conveying the idea to the other, attempting to change belief, and so forth." Finally they assume that a situated conceptualization like *convince* varies according to the situation in which it is experienced, exhibiting family resemblance not a conceptual core: "For *convince*, different situated conceptualization's represent convincing a friend, parent, policeman, mugger, audience, and so forth." Importantly they assume that emotion concepts also refer to entire situations, and thereby represent "settings, agents, objects, actions, events, interoceptions, and mentalizing" (1107). *Fear* of a runner who becomes lost on a

---

[4]See also Stanley, et al. "Of great interest are behavioral studies that demonstrate variation in implicit attitudes as a function of social context and the goals and motivations of the participant (Blair, 2002). The *context of a situation* may modulate the activation of implicit attitudes without engaging regulatory mechanisms. While this has yet to be directly addressed, there is evidence that task goals can modulate the amygdala response to Black and White faces" [emphasis added].

wooden trail at dusk, for instance, differs from the *fear* of someone unprepared to give an important presentation at work. In this latter situated conceptualization, they explain, "a different set of concepts represents the situation, including *presentation, speaking, audience, supervisor,* and many others" (1108). In essence, their laboratory experiment interrogates this hypothesis that the basic emotion approach is wrong, because in fact a situated conceptualization "produces the emotion" (1110). Or to put this another way, the situation in which an emotion concept is experienced shapes how the emotion is instantiated in the brain (1120). For example, the observation that social fear activated clusters in the left dorsolateral prefrontal cortex and inferior frontal gyrus warrants the conclusion that social fear, as opposed to physical fear, requires more executive control to cope with threatening social evaluations (1124).

"Situation" in the work of Barsalou and his students and colleagues is richly developed in terms of its brain implications, and I think his work will speak to rhetoricians interested in how rhetorical situations are embodied more generally. At the same time, rhetoricians will balk at Barsalou's modular, functionalizing focus on conceptualization that renders the situation reductively in terms of situated conceptualization while relying on elliptical figures ("and so forth," "and many others")[5] to articulate a background that recedes indefinitely into what I would consider non-conceptual affordances located in the environment, including the social environment. Considering her study of race perception and attitudes, I think Phelps is also obligated to make stronger contextual claims, and that is where rhetorical inquiry is required, whether that means tapping the disciplinary knowledge base in history, critical race theory, or rhetoric more broadly.

Finally, we should think explicitly about why certain kinds of experiment (on Black and White faces, for example) appear at a particular moment in history, thus reifying the categories unintentionally, while other possible experiments are inconceivable. In this response, I have focused on the neuroscience of race because its rhetorical dimension is so terribly obvious, and some of its practitioners already gesture toward deep integration with humanistic inquiry and not just social scientific inquiry. But the same can be said for empathy in much of the work done in contemporary neuroscience: such work is not just descriptive but prescriptive by way of its mere presence and reiteration. As in the later eighteenth-century Anglo-European context for instance, and in the US abolitionist context, one message of the moment is "empathy matters," while as Phelps et al. remind us by way of rhetorically sophisticated neuroscience, "race matters" even in the post-racial assertion of its own erasure. Certainly when neuroscientists treat social phenomena such as racial fears and empathy, rhetoric is relevant at the level of data interpretation,

---

[5]This figural critique is classic rhetoric of (neuro)science as referenced by Jack and Appelbaum, and practiced expertly by Jeanne Fahnestock.

as Jack and Appelbaum explain on their way to the *rhetoric of neuroscience*.[6] I add that rhetoric is also essential at the level of experimental design in order to adequately characterize the situation of situated cognition studies.

References

Chiao, Joan Y., et al. "Cultural Specificity in Amygdala Response to Fear Faces." *Journal of Cognitive Neuroscience* 20.12 (2008): 2167–2174. Print.

Choudhury, Suparna, and Jan Slaby. *Critical Neuroscience: A Handbook of the Social and Cultural Contexts of Neuroscience*, Chichester, West Sussex: Wiley-Blackwell, 2012. Print.

Edbauer, Jenny. "Unframing Models of Public Distribution: from Rhetorical Situation to Rhetorical Ecologies." *Rhetoric Society Quarterly* 35.4 (2005): 5–24. Print.

Fahnestock, Jeanne. *Rhetorical Figures in Science*. New York: Oxford University Press, 1999. Print.

Gibson, James J. *The Ecological Approach to Visual Perception*. Boston: Houghton-Mifflin, 1979. Print.

Griffiths, Paul, and Andrea Scarantino. "Emotions in the Wild: The Situated Perspective on Emotion." *The Cambridge Handbook of Situated Cognition*. Philip Robbins, and Aydede Murat. Cambridge: Cambridge University Press, 2009. 437–453. Print.

Gross, Daniel M. *The Secret History of Emotion: From Aristotle's Rhetoric to Modern Brain Science*. Chicago: University of Chicago Press, 2006. Print.

Gross, Daniel M. "Defending the Humanities with Charles Darwin's *The Expression of the Emotions in Man and Animals* (1872)." *Critical Inquiry* 37.1 (2010): 34–59. Print.

Heidegger, Martin. *Being and Time*. New York: Harper, 1962. Print.

Jack, Rachael E., Roberto Caldara, and Philippe G. Schyns. "Cultural Confusions Show That Facial Expressions Are Not Universal." *Current Biology* 19.18 (2009): 1543–1548. Print.

Jack, Rachael E., Roberto Caldara, and Philippe G. Schyns. "Internal Representations Reveal Cultural Diversity in Expectations of Facial Expressions of Emotion." *Journal of Experimental Psychology* 2011 [forthcoming].

Phelps, E. A., et al. "Performance on Indirect Measures of Race Evaluation Predicts Amygdala Activation." *Journal of Cognitive Neuroscience* 12.5 (2000): 729–738. Print.

Stanley, Damian, Elizabeth Phelps, and Mahzarin Banaji. "The Neural Basis of Implicit Attitudes." *Current Directions in Psychological Science* 17.2 (2008): 164–170. Print.

Preston, S. "Psychology: The Empathy Gap." *Nature London* 472.7344 (2011): 416–419. Print.

---

[6]One rather breathless example of a rhetoric of neuroscience, critiquing Rachael E. Jack et al. (2009). Rachael E. Jack et al. (2011), 19, wonder: "How could culture exert such an influence on the production and perception of basic emotion signals? Each culture embraces a specific conceptual framework of beliefs, values, and knowledge, which shapes thought and action. Culture-specific ideologies could exert powerful top-down influences on the perception of the visual environment by imposing particular cognitive styles. For example, individualistic (e.g., Western) cultures could generate tendencies to adopt local feature-processing strategies, whereas collectivist (e.g., East Asian) cultures may promote the use of global processing strategies." I respond yes they could, or maybe not. They offer a category mistake and a causal fallacy, at least. While we're at it why don't we speculate that fascist cultures could adopt *exclusive* feature-processing strategies, and evangelical cultures *transcendent* feature-processing strategies, and so on? In this case, the model of causality fits poorly (or let's say arbitrarily) with the data. Western individualism is broad enough to explain anything poorly. What we mean by culture needs to be much more precise and justifiable given the project, and this is where rhetorical sensibilities help whether one works professionally as a historian or a rhetorician.

Uexküll, Jakob. *A Foray into the Worlds of Animals and Humans: With a Theory of Meaning.* Minneapolis: University of Minnesota Press, 2010. Print.

Vygotsky, L. S., and Michael Cole. *Mind in Society: The Development of Higher Psychological Processes.* Cambridge: Harvard University Press, 1978. Print.

Wilson-Mendenhall, C. D., et al. "Grounding Emotion in Situated Conceptualization." *Neuropsychologia* 49.5 (2011): 1105–1127. Print.

Whatever Happened to the Cephalic Index? The Reality of Race and the Burden of Proof

John P. Jackson, Jr.

This article examines the work of anthropologist Franz Boas who, in the early twentieth century, argued against the existence of the stability of the cephalic index, a measure of head shape, and its relation to the mental and moral capacities of human races. The article claims that Boas successfully shifted the burden of proof to his opponents and set the stage for the scientific rejection of belief in innate racial differences in intelligence. The article urges rhetorical scholars to attend to the notions of burden of proof and presumption in scientific controversies over neurological differences.

Controversy has long been a subject of rhetorical concern (Goodnight "Science and Technology Controversy"; Ono and Sloop; Thacker and Stratman; Olson and Goodnight). Yet some controversies are difficult to examine because they are almost impossible to demarcate. How do we determine when a controversy has begun or when it has ended? The scientific controversy over innate racial differences in intelligence, for example, has been going on so long and seems so resistant to final closure that psychologist Graham Richards declared that "the controversy is not so much alive as undead" (Richards "Race" 262). The controversy over intelligence and race spanned most of the twentieth century. So long have the arguments raged over racial differences in intelligence that Richards notes the "ritualised character" of the conflict. Those claiming that there are racial differences in intelligence periodically marshal their forces against the majority who do not believe in such differences only to be beaten back. "But, as with Dracula," Richards concludes, their "allies or hidden offspring retain sufficient resources for a remake in due course and overtly tabooed racist discourse can be covertly produced as 'science'" (263).

The most visible defenders of the case for racial differences in intelligence in recent years have been Berkeley psychologist Arthur Jensen and University of Western Ontario psychologist J. Philippe Rushton. Jensen obtained national notoriety four decades ago when he resurrected the case for racial differences in IQ. Rushton's work goes far beyond the psychometric measurement of IQ and proposes a full-blown evolutionary theory that posits that intelligence is inversely correlated with sex drive and genital size. Using a terminology that harkens back to the nineteenth century, Rushton holds that "Negroids" have the highest sex drive and lowest intelligence, "Mongoloids" have the lowest sex drive and highest intelligence and "Caucasoids" range in between the two (Rushton *Race*, 244).

In a recent article Rushton and Jensen together laid out, according to their title, "thirty years of research on race differences in cognitive development." Despite the chronological limitations of their title, one section of their paper draws upon research that is over a century old. In that section, titled "Race, Brain Size, and Cognitive Ability," Jensen and Rushton refer to Samuel Morton's 1849 studies of cranial capacity, Paul Broca's 1873 work on brains weighed during autopsies, as well as studies by Franklin Bean from 1906 (255) to argue that there are racial differences in brain size, and that these differences affect intelligence. Then, leaping across a century, they claim that these long-discredited studies are supported by modern neurological research that used Magnetic Resonance Imaging to measure brain volume, as well as glucose metabolic rates to measure efficient use of energy. Jensen and Rushton's conclusion is this: "The most likely reason why larger brains are, on average, more intelligent than smaller brains is that they contain more neurons and synapses, which make them more efficient" (253).

The argument has changed little since the nineteenth century; they have merely substituted Magnetic Resonance Imaging (MRI) for Morton's birdseed. We can see how apposite Richards's characterization of the race and intelligence controversy as "undead" truly is. Jensen and Rushton move easily from the Samuel Morton's crude measurements of cranial capacity to modern studies with MRIs, but the underlying argument is the same: racial differences are real and the most important racial differences are those in intelligence. Intelligence, in turn, exists insides people's heads in the neurological structure of their brains. The very persistence of claims regarding the relationships between heads/brains, intelligence, and race suggests that to properly understand the controversy we need to look beyond the evidence itself to see what else is going on rhetorically. In order to do so, I will examine part of the controversy from nearly a century ago: the debate over the stability of the cephalic index.

The cephalic index is a ratio of the length to the breadth of the head. Once thought to be the very basis of civilization, the cephalic index is now relegated to an obscure anatomical measurement solely of concern to anatomists and, perhaps, forensic anthropologists. The question is best understood as a question in the "prehistory" of neuroscience. While neuroscience itself has a distinctive history dating back perhaps half a century (Abi-Rached and Rose), the scientific study of the brain and its relationship to our mental and moral characteristics go back much further. In this article, I will focus on one particular string of scientific premises: that people could be racially classified according to the shape of their head, that the classification related to their mental and moral capabilities, and that those capabilities were the foundation of modern civilization. I will examine how this view was critiqued by anthropologist Franz Boas (1858–1942) a century ago. Boas's arguments against the bare possibility that head shape was a stable racial marker also stood against the existence of a racial hierarchy in mental and moral characteristics.

The present article addresses an episode of a larger scientific/social controversy during what historian Elazar Barkan called "the retreat of scientific racism." Between the two world wars biologists, anthropologists, and psychologists reversed their views regarding race. From a belief in the biological reality of race and a natural racial hierarchy at the close of World War I, scientists embraced a racial egalitarianism and doubts (if not denial) about the very existence of biological race. The question is *why* the scientific communities changed their minds about these topics. In 1973 historian Will Provine argued that between the two world wars geneticists completely reversed their position on the dangers of race crossing: what was viewed as the dangers of "disharmonious crosses" was proclaimed completely harmless (794). Provine, and others following him, argued that there were no new data on, for example, the dangers of race crosses and the only reason the scientific community changed its mind on such dangers had to do with the rise of political liberalism and, especially, concerns about the Nazis after 1933. Franz Samelson argued that the rejection of racial difference in IQ was almost entirely political, there being no data to support an egalitarian interpretation of IQ scores. Elazar Barkan simply summed up two decades of historical study when he proclaimed that the main conclusion of his definitive study of the retreat was that "political beliefs had a greater impact in attitudes toward race than did scientific commitments" (343. For the contrary view see: Glass; Richards "Reconceptualizing"). The general argument here is that geneticists, anthropologists, and psychologists became racial egalitarians even though there was no evidence in support of that position. The shift to cultural explanations for human behavior over biological/racial explanations was simply "the substitution of one unproved (though strongly held) assumption for another" (Degler 187). While the evidence for scientific racism might have been lacking, there was no evidence in favor of scientific egalitarianism either. This historical claim has been used by those few scientists who maintain that race is a real biological entity and that there are qualitative differences between racial groups. These scientists bolster their case for racial differences by arguing that the defeat of their position before World War II was "political" rather than "scientific" (Lynn *Eugenics* 37–39; Lynn *The Science* 540; Pearson; Rushton "The Pioneer Fund" 243). The historical controversy thus has political and scientific implications for us today.

I examine only a slice of this larger controversy on the retreat of scientific racism: anthropologist Franz Boas's work on the existence of the cephalic index. Although Boas's work is nearly a century old, such a consideration should not be considered antiquarian as the controversy over the cephalic index remains alive, with a spate of anthropologists recently reanalyzing Boas's data to question his evidence (Gravlee, Bernard, and Leonard; Holloway; Sparks and Jantz). Few data sets are so groundbreaking that they undergo a reanalysis ninety years after being generated.

This article, then, will proceed as follows: first, I will offer an outline of how presumption and burden of proof can function as critical tools in the analysis of this scientific controversy. Second, I will sketch the idea that head shape was tied to both racial identification and neurological function. Third, I will offer a new reading of Boas's arguments about head shape that views it as seizing presumption and forcing his opponents to shoulder the burden of proof.

Burden of Proof and Arguments from Ignorance

Franz Boas captured the central problematic I am trying to frame in this article in a forward he wrote to NAACP founder Mary White Ovington's book *Half a Man*, where he wrote: "Many students of anthropology recognize that no proof can be given of any material inferiority of the Negro race; that without doubt the bulk of the individuals composing the race are equal in mental aptitude to the bulk of our own people" (vii). The problem is that while the first phrase may be true, the absence of proof of Negro inferiority is not, in itself, proof of racial equality. As we have seen, Degler and other historians charge that Boas and other scientists committed the fallacy of the *argument from ignorance*: a proposition is not true simply because it has not been proven false (or vice versa). Absent from our understanding of the retreat of scientific racism is a sophisticated understanding of the functions of presumption and burden of proof in controversy.

Douglas Walton has defined the "burden of proof" as "the strength of evidence ... required to persuade" (Walton "Burden" 234). Most scholars distinguish between two kinds of burdens of proof. On one level there is the straightforward notion that one who puts forth a proposition has an obligation to prove it. In the legal system this would be the "burden of going forward." Here, I am more interested in the burden that operates on a more global level where the term can refer to which party in the argument "carries the consequences of any residual uncertainty" (Hahn and Oaksford 42). In the law, this is often called the "risk of non-persuasion." I agree with Richard Gaskins, who argues that the concept of burden of proof, understood as the risk of non-persuasion, should be "the focal point for a much wider rhetorical function" (4). Since few scholars would maintain that uncertainty can be completely eliminated, indeed some rhetorical scholars would maintain that it is *undesirable* to seek the complete elimination of uncertainty (Scott), the concept of burden of proof would seem to be vital to our understanding of how we rhetorically negotiate our way through an uncertain world.

Evidence does not exist in a vacuum but can only be understood against a backdrop of expectations about how much evidence should be required to prove the truth (or falsity) of a given proposition. Richard Gaskins takes to task those who merely proclaim that the evidence in controversy "proves" while ignoring the larger context in which the evidence is evaluated: "The language of proof could misleadingly suggest that factual evidence alone marks the difference between ignorance and knowledge in public discussion" when in many "wider rhetorical contexts . . . the basic norms of rules for weighing evidence often contribute more fundamentally to the condition of ignorance" (4–5). Since, as David Zarefsky argues, "the need to make choices when not everything can be known is the defining feature of the rhetorical situation" ("Argumentation" 20), examining how the burden of proof functions rhetorically should help rhetorical scholars better understand the choices made by a scientific community lacking definitive evidence of racial differences. The question is not: What does the evidence prove about race?, but rather, How should we act in the face of inconclusive evidence? The answer to that question rests, in large part, on which side of the controversy has presumption and which side has the burden of proof. Several critical tools from the literature on presumption and burden of proof can illuminate Boas's case against race differences.

Presumption and Burden of Proof are Rhetorical Constructs

The concept of "burden of proof" is familiar in legal settings (where defendants are presumed innocent until the prosecution proves them guilty) and often set by specific criteria that determine the strength of the burden (which is why O.J. Simpson can be both innocent of murdering his wife and culpable for her wrongful death). Work in informal logic, philosophy of language, and rhetorical argumentation has extended the concept of burden of proof to other kinds of argumentative situations. When he imported the concepts of presumption and burden of proof from jurisprudence into rhetorical theory, Richard Whately told his readers that if they forget to claim presumption for their own position, they "may appear to be making a feeble attack, instead of a triumphant defence" (*Elements* 107). Whately described presumption, not in terms of probabilities, but merely as a "*preoccupation* of the ground, as implies that it must stand good till some sufficient reason is adduced against it; in short, that the *burden of proof* lies on the side of him who would dispute it" (*Elements* 105).

Whately's development of these concepts is reflected in another book of his, *Historic Doubts Relative to Napolean Bonaparte*, which targets Hume's critique of miracles. In a marvelous *reductio ad absurdum*, Whately argues that if we were to take Hume's critique seriously, then we would cease believing in Napoleon as surely as we ceased believing in the Resurrection. Whately developed his concepts of presumption and burden of proof as cautionary principles carrying the warning that "boundless skepticism is just as problematic as boundless credulity" (Einhorn 285). Thus, Whately's most extensive example in *Elements of Rhetoric* was that Christianity should have presumption and it was necessary for its opponents to shoulder the burden of proof, nicely reversing Hume's argument against religion.

Christianity enjoyed presumption, according to Whately, because it was an existing institution and there was a "presumption in favour of existing institutions" (*Elements* 108). Indeed, presumption could be applied to any argumentative situation because there was a presumption "against *paradoxical*, i.e., contrary to the prevailing opinion... since men are not expected to abandon the prevailing belief till some reason is shown" (109). Whately's emphasis on the presumption of the *status quo* led some commentators to criticize the concept as inherently conservative (Goodnight "The Liberal" 312–314). Gaskins, for example, reads Whately as almost anti-rhetorical when compared with writers such as Perelman who embraced the "creative, flexible, and liberating aspects of rhetoric" (34). Whately's view of presumption and burden of proof serves "openly conservative functions" and therefore "turns the wrong direction" (34).

More careful readings of Whately, however, belie such a reading (Sproule). Whately, after all, was advising speakers to claim presumption for themselves, which makes little sense if presumption is automatically assigned to one side of a dispute. Whately offered several rhetorical strategies for speakers to claim presumption. The first was to recognize that audiences may assign presumption differently in different situations; as Whately puts it, "in any one question the Presumption will often be found to lie on different sides in respect of different parties" (*Elements* 113); after all, Presbyterians will give presumption to that Church and Anglicans to their own. Second, "a Presumption may be *rebutted* by an opposite Presumption," explains Whately:

> so as to shift the Burden of Proof to the other side. E.G. Suppose you had advised the removal of some *existing* restriction: you might be, in the first instance, called on to take the Burden of Proof and allege your reasons for the change, on the ground that there is a Presumption against every Change. But you might fairly reply, "true, but there is another Presumption which rebuts the former; every *Restriction* is in itself an evil; and therefore there is a Presumption in favour of its removal, unless it can be shewn necessary for prevention of some greater evil: I am not bound to allege any *specific* inconvenience; if the restriction is *unnecessary*, that is reason enough for its abolition: its defenders therefore are fairly called on to prove its necessity." (*Elements* 113–114)

Whately does indeed give presumption to the *status quo*, but he also opens possibilities for rhetors to argue just how presumption and the burden of proof should be assigned by an audience in a particular dispute; what part of the *status quo*, for example, as relevant for the assignment of presumption and burden of proof? This was a question a speaker could enroll in his or her cause. Scholars who argue that Whately's notions are inherently conservative overlook that for Whately, the concepts were rhetorical strategies that could be employed by speakers differently in different situations.

The concepts of burden of proof and presumption have generated a lot of writing since Whately's time. Most often, this literature is generated by philosophers interested in developing normative models of argument rather than rhetorical strategies that speakers could use to persuade audiences, as Whately discovered. As critics, we can read texts for these strategies, such as placing a counter-presumption against a well-understood presumption.

Burden of Proof in Scientific Controversy

Many argumentation theorists, most famously Stephen Toulmin and Chaim Perelman, follow Whately in borrowing concepts from jurisprudence and applying them to argumentation more generally. A second wellspring of argumentation theory has been academic debate, which, throughout the twentieth century, came to been seen as a kind of laboratory for testing novel argumentative ideas, particularly in the United States (Zarefsky "Argumentation"). In both these arenas the allocation of the burden of proof is well understood. Even so, recent scholarship has questioned whether jurisprudence and academic debate are really the best exemplars for fashioning a general theory of the burden of proof. Borrowing from legal reasoning has met sharp criticism in other quarters. Gaskins, for example, charges that Toulmin lacked a basic understanding of how legal processes actually operated, especially regarding the assignment of the burden of proof (16–21). Tim Dare and Justine Kingsbury similarly argue that the pedagogical goals of academic debate where we are concerned with "the skills of team members noticing and responding to the flaws and strengths of the arguments actually presented (as opposed to the best argument that could be presented)" (506–507) are a more appropriate model to extend to other argumentative situations.

Dare and Kingsbury note that in both a court trial and an academic debate the argument ends in a binary choice: defendants are either found guilty or not; one team wins the debate the other loses. Given that the choice is forced at the end of the argumentative process in these activities, a differential allocation of argumentative burdens is a good way to ensure that a choice can, indeed, be made. Such activities have a number of goals for argument besides discovering the truth: debate has pedagogical ends, and a trial has a goal of keeping social order. While a differential allocation of argumentative burdens makes sense in situations where such a choice is required, the search for scientific truth does not face such a choice (Dare and Kingsbury 505). Hahn and Oaksford argue, "A burden of proof arises naturally wherever a putative action imposes a binary outcome (e.g., 'do' or 'do not do', 'action A or action B') that has to be reconciled with graded degrees of conviction. This gives rise to a threshold degree of conviction above which an action should be adopted, given the particular utilities associated with this action"

(57). By contrast, "Science . . . has no rules for terminating debate. Theories and results can always be subjected to further scrutiny. Furthermore, scientific arguments typically produce considerably less than certainty and science itself sits comfortably with degrees of conviction" (Hahn and Oaksford 48).

These normative claims about the nature of scientific inquiry depend on what the authors admit is an idealized picture of scientific inquiry where the only goal of the inquiry is "truth." As we will see, however, the kind of epistemic inquiry that forms the centerpiece of my story here was enmeshed in political and social controversy, the very kind of argumentative situation where a differential allocation of argumentative burdens is warranted. Indeed, we can question if the picture of science as free and unfettered inquiry into truth should be the basis of our normative models of science (Proctor 11). The idealized picture of dispassionate scientific inquiry into truth that could wait decades for a definitive answer was certainly not how Boas pictured his work. He argued that scientific inquiry into the stability of racial types was necessitated by the need for informed social policy, particularly around the subject of immigration: while the search for truth was paramount, the need for an answer was urgent. In a 1909 address to the American Association for the Advancement of Science, Boas argued this very point: "Under the pressure of these events, we seem to be called upon to formulate definite answers to questions that require the most painstaking and unbiased investigation. The more urgent the demand for final conclusions, the more needed is a critical examination of the phenomena and of the available methods of solution" ("Race Problems" 839). A rhetorical reading of the controversy, I maintain, would require seeking out how an arguer can rhetorically construct a binary choice which creates the grounds for claiming presumption for one's side, even when that controversy is conducted in the language of scientific inquiry.

Burden of Proof and the Argument from Ignorance

Walton has written extensively on his "pragmatic theory of fallacy." In the pragmatic theory, a fallacy is not always seen as an error in reasoning but as an argumentative tactic that has the semblance of good argument but is actually an argument that interferes with the collaborative goals of the arguers. On this view, a critic cannot merely label an argumentative move as fallacious because of the form the argument takes; rather the critic must have an understanding of the social context in which the argument is offered. Walton has done extensive work on many common fallacies to show how sometimes a particular argument is fallacious and sometimes it is not, depending on the social circumstances.

In his work on the argument from ignorance, Walton argues that a nonfallacious argument from ignorance employs a "negative logic," which:

> has the form: A is not proven true (false), therefore A may be presumed to be false (true). This "flip flop" type of reasoning is, as shown above, characteristic

of how presumptive reasoning functions in a context of dialogue by reversing the roles (probative obligations) of the proponent and respondent. (Walton "Nonfallacious" 385)

According to Walton, one way the argument from ignorance would be non-fallacious is if the arguer advancing it has made a reasonable effort to ensure that the knowledge base is complete if the arguers have reached "epistemic closure," often stated with the phrase, "if it were true, I would know it" ("Non-fallacious" 382). In this version, the argument from ignorance is fallacious when an arguer deceptively argues that a good-faith effort has been made to complete the knowledge base but, in fact, has not made such an effort. Conversely, "once a knowledge base is definitely closed, then if a proposition does not appear in it, we can conclude that this proposition is false" ("Nonfallacious" 182). An arguer can put forth evidence that the knowledge base is complete (or reasonably complete) and conclude that, since a proposition is not proven by the available knowledge, it is reasonable to conclude that the proposition is false. David Zarefsky concludes that the argument from ignorance:

> is considered fallacious when it prematurely closes dialogue, when it is a substi-
> tute for critical discussion. This occurs when superficial or perfunctory deliber-
> ation is treated as if it were thorough and systematic. One is far less justified in
> claiming that what is not found does not exist when one has not looked very
> hard. Or the dialogue is closed when the conclusion is universalized, treated
> not as a presumption but as an unvarying truth. In that case it is a substitute
> for, not the outcome of, careful deliberation. ("Making the Case" 289)

A rhetor can then be said to legitimately shift the burden of proof to the opponent only if the rhetor simultaneously makes a good-faith effort to find positive evidence and to urge that the search continues. Until such time that positive evidence is found, however, it is reasonable to accept the absence of evidence as the best we can do for now and act accordingly.

Boas conducted an extensive study searching for the stability of the cephalic index. After the study failed to produce such evidence, Boas argued that it was reasonable to conclude that the cephalic index could not be used in racial classi-fication and thus that the mental attributes supposedly based on head shape were similarly illusory. In the larger controversy about racial equality, Boas did not argue that there is evidence for racial equality; rather, he argued that he made a good-faith effort to find evidence of racial inequality and has failed to find any. Hence the burden of proof should fall on those who maintain white supremacy.

Neurology, Head Shape, and Race: Neurology's Pre-History

Linking personality and character traits to skull shape represented the very best thinking in nineteenth-century neuroanatomy. Franz Joseph Gall's (1758–1828) argument that the brain could be divided up into locations and that each location of the brain controlled different physiological functions is still seen as an important precursor to today's neuroscience (Clarke and Jacyna 220–241). Charles Darwin's biographers note that in his medical training at Edinburgh, his professors stressed the relationship between head shape and mental function, which shows that such beliefs had a "serious neuroanatomical side" (Desmond and Moore 34).

The goal of this neuroanatomical work, much of which went under the umbrella term "craniometry," was to work out the laws that governed the relationship between skull shape and intricate mental processes. In all of these studies, the size and shape of the head was a proxy for the brain, which was, in turn, a proxy for a capacity for intelligence and moral characteristics. Far from an occupation for a few medical/scientific quacks, the scientific attempt to link the size and shape of the head to intelligence dominated nineteenth-century anthropology. The work set the stage for twentieth-century IQ researchers who embrace the central tenants of craniometry, namely "that intelligence could be unitary descriptor of native intellectual endowment that varied by degrees" and "the possibility of precise, quantitative measurement of those degrees" (Carson 109).

The central rhetorical tactic of craniometry was a coexistential argument that connected "the essence and its manifestations" (Perelman and Olbrechts-Tyteca 293). The argument moved from an attribute that could be observed (skull shape) to an underlying essence that cannot be observed (mental or moral characteristics). The same coexistential argument can be found as a central component to racist ideology, which has as one of its central tenets that the physical appearance of an individual indicated their mental capacity and moral character (Smedley 28). In the United States, for rather obvious reasons, we tend to assume that those outer racial characteristics were always based on skin color. However, at the end of the nineteenth century and the beginning of the twentieth century it was head shape, not skin color, that was the central focus of scientific investigation of race. This was most obvious in the Nazi regime where craniometry formed the basis of the anthropological work of Hans F.K. Guenther who, with Alfred Rosenberg, was the chief architect of Nazi racial ideology. Guenther's obsession with measuring heads is largely ignored by histories of Nazi Germany, leading Jennifer Michael Hecht to declare that "Our failure to notice the numerical basis of race theory in this period is bizarre, considering how clearly such [skull] measurements dominate the literature of the time" ("Vacher de Lapouge" 300).

Even earlier than the Nazis, however, in the nineteen-teens and -twenties the United States was receiving waves of immigrants from southern and eastern Europe. While obviously possessing white skin, many "native Americans" considered these new immigrants to be racially—that is to say *biologically*—inferior. It took several generations for these new immigrants to be assimilated into a generic "white" race (Jacobson; Roediger).

While markers like skin color or eye shape might serve for identifying the three "great races" (Caucasoid, Negroid, Mongoloid), for distinguishing between the "European races" other measures were necessary. For many scientists, a key marker of race was the cephalic index: a ratio of the width of the head to the length of the head invented/discovered by Swedish anatomist Andres Retzius (1796–1860). The actual formula was:

$$CI = \frac{\text{Width of the head}}{\text{Length of the head}} \times 100$$

Which yielded the following racial categories:

I. dolichocephalic: $CI = 74.99$ and down.
II. mesocephalic: 75–79.9
III. brachycephalic: 80–up

"The shape of the human head," concluded William Z. Ripley in 1899, " by which we mean the general proportions of the length, breadth, and height . . . is one of the best available tests of race known" (37). Through the cephalic index (and other measures), Ripley mapped three European races: The dolichocephalic Nordics, the mesocephalic Alpines, and the brachycephalic Mediterraneans. Only the Nordic, sometimes called Aryan, race was capable of creative civilization. The key was the shape of the head, as Jean Finot, writing in the popular periodical *Contemporary Review* in 1911 noted that "our distinction of races" almost completely owed to "Brain measurement and its sister art, head measurement" (480).

The cephalic index was embraced by physical anthropologists and others seeking to establish racial typologies because it was thought to be immune from environmental influence and hence immune from natural selection. Darwin had taught that species were not fixed types but were mutable over time. The cephalic index, however, was not subject to selection pressure that brought about the changes in morphology that constituted evolution. Yale geographer Ellsworth Huntington sums up the argument thus:

> So far as we are aware the shape of people's heads cannot be influenced by their food, their occupation, or their social and economic conditions. Nor can we see how climatic selection could weed out one type of head as it weeds out one type of complexion. In fact, the shape of the head has been supposed to be one of the most stable racial characteristics and to be one of the most invariable features of man's physical inheritance. (172–173)

The cephalic index, then, was seen as a glimpse into the primordial past when the true racial types were constituted. In this way, scientists attempted to hold onto the stability of racial types in the face of a Darwinian science that would otherwise require them to surrender such fixed and static categories. It was this belief that allowed America's most notable racist, Madison Grant, to declare his allegiance to Darwinian principles while decrying the "widespread and fatuous belief in the power of environment, as well as of education and opportunity to alter heredity, which arises from the dogma of the brotherhood of man, derived in turn from the loose thinkers of the French Revolution and their American mimics" (Grant 14).

Scientists used the cephalic index for more than establishing the reality of racial types; it could also be used to argue for the existence of a racial hierarchy. Among the three European races: the long-headed Nordic/Aryan, the mid-range Alpine, and the round-headed Mediterranean, many believed only the Nordics were natural leaders because they were energetic and creative. The lesser races were perhaps good workers, but lacked the spark that made the Nordic the drivers of civilization: "In the hierarchy of races the first place must be given to *Homo Europus* (the dolichocephalic-blond or so-called Aryan), while *Homo Alpinus* (the brachycephalic type) and the Mediterranean probably rank in the order named" (Lapouge and Closson 60). Unless something was done to check the widespread interbreeding of the noble Nordic with inferior racial types, anthroposociologist Vacher de Lapouge warned, "In the next century people will be slaughtered by the millions for the sake of one or two degrees on the cephalic index" (quoted in Hecht, *End of the Soul* 168).

The blustering assurance of racists like Lapouge and Grant masked the fact that there was little consensus among scientists as to the basis of racial classification. While the cephalic index was highly regarded, if only for its ease of computation, scientists recognized that the index was best when it could be correlated with other morphological traits. However, the more scientists searched for a stable set of physical markers that could reliably be used for racial taxonomy, the more elusive such markers appeared to be. Historian Jonathan Spiro gives a glimpse at what he calls the "classificatory swamp" of early twentieth century race researchers:

> Cuvier claimed there were three human races, Saint-Hilaire thought there were four, Quatrefages five, Virey six, Peschel seven, Agassiz eight, Flower eleven, Mueller twelve, Saint-Vincent fifteen, Desmoulins sixteen, Topinard nineteen, Morton twenty-two, Broca twenty-seven, Deniker twenty-nine, Haeckel thirty-four, Crawford sixty, Burke sixty-three—and so on. (102)

Despite this lack of consensus, there was little scientific doubt that race was a real scientific category and the loud voices proclaiming the reality of racial differences continued. Against this backdrop, Franz Boas conducted a large-scale study to measure the stability of racial differences in head shape.

Franz Boas

Rightly considered the founder of American cultural anthropology, Boas was a fierce critic of racial typologies both in his professional and popular work; indeed historian Thomas Gossett claims that "Boas did more to combat race prejudice than any other person in history" (418). By separating "culture" from "biology," Boas brought a new explanatory framework to the understanding of human history and experience. He also brought a new rigor to both cultural and physical anthropology, demanding precision in claims and evidence that were not typical of anthropological work at the time.

The United States was undergoing an extensive debate about immigration into the United States at the end of the nineteenth and beginning of the twentieth centuries. These immigrants, overwhelmingly from southern and eastern Europe, were considered racially inferior and eventually led to increasingly draconian immigration restriction legislation, culminating with the Johnson Act of 1924, which effectively ended immigration to the United States from countries that were not of pure Nordic stock (Spiro 198–233). As a prelude to legislation, the U.S. Congress commissioned a number of studies of immigration and immigrants. Boas's work on head shape was one of these forty-two thick volumes produced for the U.S. Immigration Commission in 1910 and that Boas brought to a wider audience with his book, *The Mind of Primitive Man* (Barkan 82–85; Stocking 161–194). In these works, Boas carefully distinguished between what evidence was available to researchers and inferences that one could draw from that evidence. He also employed rhetorical strategies that used presumption and burden of proof to his rhetorical advantage: He opposed presumptions with opposing presumptions. He rhetorically constructed a binary choice. He argued that while more work needed to be done, it was rational to reject the stability of race and racial differences. Finally, he refused to shut down the debate and constantly called for further research, thus avoiding a fallacious argument from ignorance.

Opposing Presumption with an Opposing Presumption

Boas rejected the notion that the stability of head form was something anthropologists had truly proven. He argued that "The general tendency of anthropological inquiry has been to assume the permanence of the anatomical characteristics of the present races, beginning with the European races of the early neolithic times" (*Mind* 44). Thus, Boas attempted to refute the presumption granted the stability of head form by calling it an assumption, rather than an established fact. Moreover he placed it against a different presumption, that of modern Darwinian theory, which denied any sort of stability to organic form: "The principles of biological science forbid us to assume a permanent stability of bodily form. Our whole modern concept of the development of varieties and of species is based on the assumption of cumulative or sudden variation. The variations that have been found in the human body are quite in accordance with this view" (*Mind* 41). The upshot of the variety of biological phenomena led Boas to completely reject the very notion of biological types. "Since all biological phenomena are variable phenomena," Boas wrote, "The fact that anthropologists are in the habit of calling heads of a length/breadth index of 80 and more, brachycephalic heads, does not constitute brachycephaly a distinct biological type, but is a mere convenience of description" ("Changes" 542).

Despite his notion that there could be no biological types whatsoever, Boas was aware that numerous studies had, in fact, shown marked distributions of head form, and so it *appeared* as if these types existed. He admitted that "All studies of the distribution of head-forms and of other anthropometric traits have shown uniformity over considerable continuous areas and through long periods" (*Mind* 45). However, he noted that the explanation for that stability was not in evidence, but was an inference from the evidence. He claimed that it was only a "natural inference... that heredity controls anthropometric forms, and that these are therefore stable" (*Mind* 44–45). Since Boas thought that typologies were at best useful ways scientists could categorize their objects of study, and at worst dangerous, unscientific illusions, it was this assumption and the inference from it that Boas tested in his study.

Rhetorically Constructing a Binary Choice

Boas's target in the Immigration Commission study was the crown jewel of racialist anthropology: the cephalic index (CI). Even before the immigrant study, Boas had published on the cephalic index, noting it was not fixed characteristic but one that changed as an individual aged, for example (Boas "Form of the Head"). In one study of 282 Sioux Indians he played the CI against another favorite racial trait—cranial capacity—and found there to be no statistical relationship (Boas "Cephalic Index").

Boas argued there were two possible explanations for why heads had the shapes they did: (1) "selection," by which he meant genetically caused traits that would accumulate over generations because of differential death rates or (2) "environment," which could include such factors as diet, swaddling techniques of infants and the like. On the surface, either theory was equally plausible, but each would require evidence in order to warrant scientific assent. He framed the issues this way:

> It goes without saying that haphazard application of unproved though possible theories cannot serve as proof of the effectiveness of selection or of environment in modifying types. The effectiveness of selection can be proved only by an investigation of the surviving members of a type as compared to those eliminated by death, or of a shifting of population connected with the selection of a certain type. The influence of environment requires the direct comparison of parents living under one environment with children living under another environment. I cannot give any example in which the influence of selection has been proved beyond cavil. (*Mind* 52–53)

The immigrant study provided Boas with an opportunity to conduct a study of the second sort. As Elazar Barkan describes the study, "Boas examined the skull measurements of first-generation Americans of Italian and Jewish descent, and compared their cephalic index to that of the populations in the country of origin. He concluded that the differences between first- and second-generation Americans born of different ethnicities was smaller than between the respective European populations" (Barkan 83). Rather than finding the stability in head form that allowed anthropologists to classify European races, Boas found that within a single generation the head shape changed significantly:

> In most of the European types that have been investigated the head form, which has always been considered one of the most stable and permanent characteristics of human races, undergoes far reaching changes coincident with the transfer of the people from European to American soil. For instance, the east European Hebrew, who has a very round head, becomes more long-headed; the south Italian, who in Italy has an exceedingly long head, becomes more short-headed; so that in this country both approach a uniform type, as far as the roundness of the head is concerned. (Boas *Changes* 5)

Boas argued that, since one generation was far too short a time for any genetic change to have caused his results, "We are therefore compelled to draw the conclusion that if these traits change under the influence of environment, presumably none of the characteristics of the human types that come to America remain stable. The adaptability of the immigrant seems to be very much greater than we had a right to suppose before our investigations were instituted" (Boas *Changes* 2).

Boas constructed a choice for his audience: one could either believe that head shape was caused by selection or by environmental pressures. His study had shown that head shape changed to rapidly to be accounted for by selection, however he did not claim to have settled the question. In further studies, however, Boas believed the burden was on those who continued to believe in the stability of head shape.

Rhetorically Shifting the Burden of Proof

Boas claimed to have proven that head shape was plastic; however, even his subsequent defenses of his study refused to posit any specific underlying environmental causes. His only claim was that he had eliminated genetic causes: "It will, therefore, be seen," he wrote, "that my position is that I find myself unable to give an explanation of the phenomena, and that all I try to do is to prove that certain explanations are impossible. I think this position is not surprising, since what happens here happens in every purely statistical investigation. The resultant figures are merely descriptions of facts which in most cases cannot be discovered by any other means" (Boas "Changes" 555–556). That he could not name specific causes for the results he obtained was not immediately relevant for Boas, because what he felt he had accomplished was to eliminate the possibility that such changes could be accounted for by the strictly biological methods available via natural selection. If one assumed that head shape was stable, then the changes Boas had discovered could only be explained through the differential death rates of immigrants in relation to foreign populations. While not dismissing the possibility outright, he did note that it was extremely unlikely and put the burden of proof firmly on the shoulders of those who would maintain such a view:

> Earnest advocates of the theory of selection might claim that all these changes are due to the effects of changes in death-rate among foreign-born and American-born; that either abroad or here individuals of certain types are more liable to die, and that thus these changes are gradually brought about. On the whole, it seems to my mind, the burden of proof would lie entirely on those who claim such a correlation between head-index, width of face, etc., and death-rate,—a correlation which I think is highly improbable, and which could be proposed only to sustain the theory of selection, not on account of any available facts. (Boas *Mind* 63)

Because there was no stable head shape, Boas conjectured further that there could be no differences in mental capacities among human types, to the extent that those capacities were indicated by head shape. In his report for Congress, he made the call to shift evidential burdens quite explicit:

> Whatever the extent of these bodily changes may be, if we grant the correctness of our inferences in regard to the plasticity of human types, we are necessarily led to grant also a great plasticity of the mental make-up of human types. . . . It is true that this is a conclusion by inference; but if we have succeeded in proving changes in the form of the body, the burden of proof will rest on those who, notwithstanding those changes, continue to claim the absolute permanence of other forms and functions of the body. (Boas *Changes* 76)

Here, Boas carefully laid out what he believed he had shown and what kind of work needed to be done. He did not claim to have settled the question regarding the stability of head shape, nor to show that all races were equal in mental ability. He did, however, spell out what he thought needed to be done in order for someone to claim that different races had different abilities. Far from calling to end research on the question he urged others to take up the task.

Calling for Further Research

When Boas took these findings in his popular book, *The Mind of Primitive Man*, he offered a reinterpretation of the "racial basis of civilization" by separating out cultural achievement from the biological trappings represented by race. Boas contradicted the widespread belief that the highest civilizations of the world were European and that there was a greater mental aptitude among Europeans than among other peoples. "Before accepting this conclusion which places the stamp of eternal inferiority upon whole races of man," Boas notes, "we may well pause, and subject the basis of our opinions regarding the aptitude of different peoples and races to a searching analysis" (*Mind* 2). Boas then couples his critique of fixed, anatomical differences among racial types with a whirlwind tour of world history that notes the contributions of non-European people to the culture of Europe before concluding:

> We have found that the unproved assumption of identity of cultural achievement and of mental ability is founded on an error of judgment; that the variations in cultural development can as well be explained by a consideration of the general course of historical events without recourse to the theory of material differences of mental faculty in different races. We have found, furthermore, that a similar error underlies the common assumption that the white race represents physically the highest type of man, but that anatomical and physiological considerations do not support these views. (Boas *Mind* 29)

Boas coupled the lack of evidence for the racial interpretation with a plausible cultural interpretation for what was widely believed to be biological differences in achievement. Yet he was unwilling to claim that he had provided the complete solution to the vexed problem of racial differences. "I repeat that I have no solution to offer," he concluded, "I have only stated the results of my observations and considered the plausibilities of various explanations that suggest themselves, none of which were found satisfactory. Let us await further evidence before committing ourselves to theories that cannot be proven" (Boas "Changes" 562). Boas's calls for further research were always tied to a conclusion that racial equality must be the default position. The last chapter of *Mind*, "Race Problems in the United States," addresses "the Negro problem." Here, he claimed

> When the bulky literature of this subject is carefully sifted, little remains that will endure serious criticism; and I do not believe that I claim too much when I say that the whole work on this subject remains to be done. The development of modern methods of research makes it certain that by careful inquiry definite answers to our problems may be found. Is it not, then, our plain duty to inform ourselves, that, so far as that can be done, deliberate consideration of observations may take the place of heated discussion of beliefs in matters that concern not only ourselves, but also the welfare of millions of negroes. (277–278)

Boas understood that the study of racial differences in the United States could not be pure and unfettered inquiry into a strictly scientific question. There were important political questions that would follow from these scientific conclusions. One way to deal with the tensions was to fix the burden of proof on his opponents while simultaneously calling for further research into racial differences. By so doing, Boas could avoid charges that he was closing off empirical research while avoiding the implementation of racist social policies.

Conclusion

Boas's revolutionary finding was fiercely criticized at the time (e.g., Radosavljevich; Hooton) leading Boas to publish all of his raw data in 1928. The data, amounting to over five hundred pages of statistical tables, were prefaced with Boas's remark that "It seemed necessary to make the data accessible because a great many questions relating to heredity and environmental influences may be treated by means of this material" (Boas *Materials* viii). Publishing his raw data allowed researchers to reanalyze his results in the early twenty-first century. Yet, the publication of the raw data was also a deft rhetorical move by Boas, for it made clear that he was not foreclosing questioning of his position. Indeed, he was so interested in pursuing a more satisfactory answer that he made his data available to other researchers who were interested in finding the answer to the question of the stability of race and the reality of racial differences. That the data were then used, nine decades later, to explore once again the question of the stability of cephalic index is testimony to the magnitude of Boas's accomplishment. In their reanalysis, Clarence Gravlee, H. Russell Bernard, and William Leonard reaffirm Boas's conclusion about the plasticity of head form and conclude "Given the prevailing faith in the absolute permanence of cranial form, Boas's demonstration of change—any change—in the cephalic index within a single generation was nothing short of revolutionary" (136). Boas issued a challenge to those who maintained the reality of race and neurological differences between races to produce evidence in support of those positions.

A century later, however, a few scientists, such as Rushton and Jensen, carry on the fight for neuroanatomical differences in brain structure that led to differences in intelligence. One reason why this undead controversy refuses to stay buried is that even defenders of racial egalitarianism recognize, "It is, of course, logically impossible to prove that there are no fundamental genetically based differences in behaviour among human groups and races; that would amount to proving the null hypothesis. Propositions of human equality, therefore, always remain fragile and vulnerable to any who care to challenge them" (Weizmann *et al.* 11). For their part, Jensen and Rushton quickly demand that their opponents produce just such a proof. "We believe," they conclude, "the burden of proof must shift to those who argue for a 100% culture-only position" (279). Rhetorical scholarship can perhaps help close this lumbering controversy by calling attention to these kinds of rhetorical moves. That is, controversy is not only about evidence, but also about claims about who has the obligation to produce evidence.

A century after Boas conducted his study, scientific claims about our neurological capacities and potentials still have consequences for how people are treated both in daily life and in the political sphere. Indeed, some claim that modern neuroscience can answer ancient philosophical questions about what it means to be human. Yet, as historian Roger Smith reminds us, such an empirical claim about human nature "fundamentally, and irreducibly, concerns meaning, values, social rules, and the expressive world made possible by language" (Smith 110). Rhetoricians, attuned to how humans construct arguments and the social rules for how empirical evidence is enrolled in social controversies, are particularly well-positioned to enlighten us about neuroscience's claims to tell us who we are.

References

Abi-Rached, J. M., and N. Rose. "The Birth of the Neuromolecular Gaze." *History of the Human Sciences* 23.1 (2010): 11–36. Print.

Barkan, Elazar. *The Retreat of Scientific Racism: Changing Concepts of Race in the Britain and the United States between the World Wars.* Cambridge: Cambridge University Press, 1992. Print.

Boas, Franz. "The Cephalic Index." *American Anthropologist* 1.3 (1899): 448–461. Print.

———. "The Form of the Head as Influenced by Growth." *Science* 4 (1896): 50–51. Print.

———. "Race Problems in America." *Science* 29 (1909): 839–849. Print.

———. "Changes in the Bodily Form of Descendants of Immigrants." *American Anthropologist* 14.3 (1912): 530–562. Print.

———. *Changes in the Bodily Form of Descendants of Immigrants.* Senate Document 208 61st Congress, 2nd Session sess., 1911. Print.

———. "Forward." *Half a Man: The Status of the Negro in New York,* by Mary White Ovington, vii–ix. New York: Longmans, Green, and Co., 1911. Print.

———. *Materials for the Study of Inheritance in Man.* New York: Columbia University Press, 1928. Print.

———. *The Mind of Primitive Man.* New York: MacMillan, 1911. Print.

Carson, John. *The Measure of Merit: Talents, Intelligence, and Inequality in the French and American Republics, 1750–1940.* Princeton, NJ: Princeton University Press, 2007. Print.

Clarke, Edwin, and L. S. Jacyna. *Nineteenth-Century Origins of Neuroscientific Concepts.* Berkeley: University of California Press, 1987. Print.

Dare, Tim, and Justine Kingsbury. "Putting the Burden of Proof in its Place: When are Differential Allocations Legitimate?" *Southern Journal of Philosophy* 46 (2008): 503–518. Print.

Degler, Carl N. *In Search of Human Nature: The Decline and Revival of Darwinism in American Social Thought.* New York: Oxford University Press, 1991. Print.

Desmond, Adrian J. and James R. Moore. *Darwin's Sacred Cause: How a Hatred of Slavery Shaped Darwin's Views on Human Evolution.* Boston: Houghton Mifflin Harcourt, 2009. Print.

Einhorn, Lois J. "Did Napoleon Live? Presumption and Burden of Proof in Richard Whately's Historic Doubt Relative to Napoleon Bonaparte." *Rhetoric Society Quarterly* 16.4 (1986): 285–297. Print.

Finot, Jean. "Long Heads, Short Heads." *Contemporary Review* 99 (1911): 479–486. Print.

Gaskins, Richard H. *Burdens of Proof in Modern Discourse.* New Haven: Yale University Press, 1992. Print.

Glass, Bentley. "Geneticists Embattled: Their Stand against Rampant Eugenics and Racism in America during the 1920s and 1930s." *Proceedings of the American Philosophical Society* 130 (1986): 130–154. Print.

Goodnight, G. Thomas. "The Liberal and the Conservative Presumptions: On Political Philosophy and the Foundation of Public Argument." *Conference Proceedings—National Communication Association/American Forensic Association (Conference on Argumentation)* (1980): 304–337. Print.

———. "Science and Technology Controversy: A Rationale for Inquiry." *Argumentation & Advocacy* 42.1 (2005): 26–29. Print.

Gossett, Thomas F. *Race: The History of an Idea in America.* Dallas: Southern Methodist University Press, 1963. Print.

Grant, Madison. *The Passing of the Great Race or the Racial Basis of European History.* New York: Scribners, 1916. Print.

Gravlee, Clarence C., H. Russell Bernard, and William R. Leonard. "Heredity, Environment, and Cranial form: A Reanalysis of Boas's Immigrant Data." *American Anthropologist* 105 (2003): 125–138. Print.

Hahn, Ulrike, and Mike Oaksford. "The Burden of Proof and its Role in Argumentation." *Argumentation* 21.1 (2007): 39–61. Print.

Hecht, Jennifer Michael. *The End of the Soul: Scientific Modernity, Atheism, and Anthropology in France.* New York: Columbia University Press, 2003. Print.

———. "Vacher De Lapouge and the Rise of Nazi Science." *Journal of the History of Ideas* 61 (2000): 285–304. Print.

Holloway, R. L. "Head to Head with Boas: Did He Err on the Plasticity of Head Form?" *Proceedings of the National Academy of Sciences of the United States of America* 99.23 (2002): 14622–14623. Print.

Hooton, Earnest Albert. *Up from the Ape.* New York: The Macmillan Company, 1931. Print.

Huntington, Ellsworth. *World-Power and Evolution.* New Haven: Yale University Press, 1919. Print.

Jacobson, Matthew Frye. *Whiteness of a Different Color: European Immigrants and the Alchemy of Race.* Cambridge: Harvard University Press, 1998. Print.

Jensen, Arthur. "How Much Can We Boost IQ and Scholastic Achievement?" *Harvard Educational Review* 3 (1969): 1–124. Print.

Lapouge, Georges Vacher De and Carlos C. Closson. "The Fundamental Laws of Anthropo-Sociology." *The Journal of Political Economy* 6.1 (1897): 54–92. Print.

Lynn, Richard. *Eugenics: A Reassessment.* Westport, Conn.: Praeger, 2001. Print.

———. *The Science of Human Diversity: A History of the Pioneer Fund.* Lanham, MD: University Press of America, 2001. Print.

Olson, Kathryn M. and G. Thomas Goodnight. "Entanglements of Consumption, Cruelty, Privacy, and Fashion: The Social Controversy Over Fur." *Quarterly Journal of Speech* 80 (1994): 249–276. Print.

Ono, Kent A., and John M. Sloop. "Critical Rhetorics of Controversy." *Western Journal of Communication* 63.4 (1999): 526–539. Print.

Pearson, Roger. *Heredity and Humanity: Race, Eugenics and Modern Science.* Washington, DC: Scott-Townsend, 1996. Print.

Perelman, Chaïm, and Lucie Olbrechts-Tyteca. *The New Rhetoric: A Treatise on Argumentation [Traité de l'argumentation. English].* Notre Dame, Ind.: University of Notre Dame Press, 1969. Print.

Proctor, Robert N. *Value Free Science? Purity and Power in Modern Knowledge.* Cambridge MA: Harvard University Press, 1991. Print.

Provine, William B. "Geneticists and the Biology of Race Crossing." *Science* 182 (1973): 790–6.

Radosavljevich, Paul R. "Professor Boas' New Theory of the Form of the Head—A Critical Contribution to School Anthropology." *American Anthropologist* 13.3 (1911): 394–436. Print.

Richards, Graham. *"Race", Racism, and Psychology: Towards a Reflexive History.* London: Routledge Press, 1997. Print.

———. "Reconceptualizing the History of Race Psychology: Thomas Russell Garth (1872–1939) and How he Changed his Mind." *Journal of the History of the Behavioral Sciences* 34 (1998): 15–32. Print.

Ripley, William Zebina. *The Races of Europe: A Sociological Study.* New York: D. Appleton and Company, 1899. Print.

Roediger, David R. *The Wages of Whiteness: Race and the Making of the American Working Class.* New York: Verso, 1991. Print.

Rushton, J. Philippe. *Race, Evolution, and Behavior: A Life History Perspective.* New Brunswick, NJ: Transaction Publishers, 1995. Print.

———. "The Pioneer Fund and the Scientific Study of Human Differences." *Albany Law Review* 66 (2002): 207–262. Print.

Rushton, J. Philippe, and Arthur R. Jensen. "Thirty Years of Research on Race Differences in Cognitive Ability." *Psychology, Public Policy, and Law* 11.2 (2005): 235–294. Print.

Samelson, Franz. "From 'Race Psychology' to 'Studies in Prejudice': Some Observations on the Thematic Reversal in Social Psychology." *Journal of the History of the Behavioral Sciences* 14 (1978): 265–278. Print.

Scott, Robert L. "On Viewing Rhetoric as Epistemic." *Central States Speech Journal* 18 (1967): 9–17. Print.

Smedley, Audrey. *Race in North America: Origin and Evolution of a Worldview.* 2nd ed. Boulder: Westview Press, 1998. Print.

Smith, Roger. *Being Human: Historical Knowledge and the Creation of Human Nature.* New York: Columbia University Press, 2007. Print.

Sparks, C. S., and R. L. Jantz. "A Reassessment of Human Cranial Plasticity: Boas Revisited." *Proceedings of the National Academy of Sciences* 99.23 (2002): 14636–14639. Print.

Spiro, Jonathan P. *Defending the Master Race: Conservation, Eugenics, and the Legacy of Madison Grant.* Burlington: University of Vermont Press, 2009. Print.

Sproule, J. M. "The Psychological Burden of Proof: On the Evolutionary Development of Richard Whately's Theory of Presumption." *Communication Monographs* 43.2 (1976): 115–129. Print.

Stocking, George W. *Race, Culture, and Evolution: Essays in the History of Anthropology.* Chicago: University of Chicago Press, 1968. Print.

Thacker, B., and J. F. Stratman. "Transmuting Common Substances: The Cold Fusion Controversy and the Rhetoric of Science." *Journal of Business and Technical Communication* 9.4 (1995): 389–424. Print.

Toulmin, Stephen. *Uses of Argument.* Cambridge: Cambridge University Press, 1958. Print.

Walton, Douglas N. "Burden of Proof." *Argumentation* 2.2 (1988): 233–254. Print.

———. "Nonfallacious Arguments from Ignorance." *American Philosophical Quarterly* 29 (1992): 381–387. Print.

———. *A Pragmatic Theory of Fallacy.* Tuscaloosa: University of Alabama Press, 1995. Print.

Weizmann, Fredric, et al. "Differential K Theory and Racial Hierarchies." *Canadian Psychology/ Psychologie Canadienne* 31.1 (1990): 1–13. Print.

Whately, Richard. *Elements of Rhetoric.* Fifth ed. London: B. Fellowes, 1836. Print.

———. *Historic Doubts Relative to Napolean Bonoparte.* Seventh ed. London: B. Fellowes, 1841. Print.

Zarefsky, David. "Argumentation in the Tradition of Speech Communication Studies." In *Perspectives and Approaches: Proceedings of the Third ISSA Conference on Argumentation*, vol. 1, ed. Frans H. Van Eemeren, Rob Grootendorst, J. Anthony Blair, and Charles A. Willard. Amsterdam: SICSAT, 1995. Print.

———. "Making the Case for War: Colin Powell at the United Nations." *Rhetoric & Public Affairs* 10.2 (2007): 275–302. Print.

The Skeleton on the Couch: The Eagleton Affair, Rhetorical Disability, and the Stigma of Mental Illness

Jenell Johnson

In 1972, vice presidential candidate Thomas Eagleton revealed to the American public that he had been hospitalized for depression on three occasions. The revelation seriously damaged the campaign of his running mate, George McGovern, and eventually led to Eagleton's dismissal from the ticket. This article seeks to understand the Eagleton Affair by showing how the stigma of mental illness functions as a form of rhetorical disability. Using a reading of stigma in ancient Greece and the work of Erving Goffman, this article argues that stigma can be viewed as a constitutive rhetorical act that also produces a disabling rhetorical effect: kakoethos, or bad character.

I never have viewed the hospitalizations in terms of being skeletons. I view skeletons as something you've done that is sinister, corrupt, evil, filthy.... You can call it nitpicking if you like. I don't. I'm not ashamed. There is nothing dirty or corrupt or evil about the fact that I had voluntarily gone into a hospital.

—Thomas Eagleton, "Eagleton's Own Odyssey"

On 31 July 1972, Missouri senator Thomas Eagleton was forced off the Democratic presidential ticket after spending just eighteen days as George McGovern's running mate. At a press conference six days earlier, Eagleton revealed to the American public that he had been hospitalized three times for depression and had received electroconvulsive therapy (ECT) on two separate occasions. The disclosure dealt a blow to McGovern's campaign from which it would never fully recover. Although McGovern initially declared that he would stand behind Eagleton "1000%," after a week of furious debate inside and outside his campaign, McGovern asked

Eagleton to resign and replaced him with Sargent Shriver, John F. Kennedy's brother-in-law. What happened next is the stuff of electoral legend. McGovern lost the 1972 election to Richard Nixon, winning only thirty-seven percent of the popular vote and a single state: Massachusetts. Eagleton would later describe his role as nothing more than a "rock" in McGovern's landslide loss. That might be true, McGovern retorted, "but landslides begin with a single rock" (McGovern 216).

The "Eagleton Affair," as it came to be known, is significant for a number of reasons. If we are to believe McGovern and a number of other commentators, the Eagleton Affair played a major role in the 1972 presidential election. Although it is unlikely that George McGovern would have defeated Richard Nixon under any set of circumstances, it is reasonable to assume that without the stain of the Eagleton Affair, the margin would not have been quite as wide. The Eagleton Affair also thrives in American political discourse as a point of comparison whenever a "skeleton" is revealed that threatens to destroy a political candidate, and particularly when that skeleton may have spent time on a psychiatrist's couch. During the 1988 presidential election, for example, Eagleton's name was invoked repeatedly after Michael Dukakis became the subject of rumors that he had received treatment to cope with the death of his brother (Rosenthal). As a political lesson, the Eagleton Affair is so rhetorically malleable that during the 2008 presidential campaign, mainstream pundits and bloggers alike made comparisons between Eagleton and *all four* major party candidates due to various skeletons or gaffes that threatened to derail their respective candidacies.[1]

As we focus attention in this special issue on the rhetorical dimensions of neurological difference, the Eagleton Affair offers an ideal case study: during the eighteen days of the Eagleton Affair, the United States engaged in a feverish public debate about mental illness and disability at a level of intensity and detail that remains unparalleled. These debates usually centered on two questions: first, was Eagleton obligated to disclose his psychiatric history to McGovern and the American people? And second, was McGovern justified in asking Eagleton to leave the ticket? In this article, I take up these questions by analyzing how stigma, a critical concept familiar to disability studies, functioned rhetorically in the Eagleton Affair, bringing attention to stigma as both rhetorically constituted and rhetorically disabling. Given the longstanding connection between rhetoric and politics, and the direct connection between rhetorical ability and political achievement, to be disabled rhetorically is ultimately to be disabled politically (cf. Lewiecki-Wilson 158; Brueggemann 11). If to be disabled mentally is to be disabled rhetorically (Prendergast 57), then mental illness might be one of the defining disabilities for political life.

---

[1] See articles by Green, Gizbert, and Wills on the parallels between Eagleton and Sarah Palin, as well as McCain, "Obama's Eagleton Affair," in which the Jeremiah Wright incident is compared to the Eagleton Affair; Ham, "Is Biden Another Eagleton?"; and Soreff and Bazemore.

Below, I first briefly define rhetorical disability not as the property of an individual rhetor, but as a failure of the rhetorical environment, a product of the conditions that grant or deny rhetors what Catherine Prendergast has termed "rhetoricability" (57). Using an interpretation of punitive stigma in the ancient world and the work of Erving Goffman, I then explore stigma as a constitutive rhetorical act that also produces a disabling rhetorical effect: what I call "*kakoethos*," or bad character. Last, I turn to the case of Thomas Eagleton in order to show how a rhetorical understanding of stigma can help us to better understand the Eagleton Affair as well as other circumstances where stigma's mark of bad character threatens rhetoricability.

Rhetorical Disability

Rhetorical ability has been traditionally understood through a perennial debate in ancient texts—is rhetorical ability the result of talent or training? Are rhetors born, or can they be made? In Cicero's "De Oratore," for example, Crassus defends natural talent by listing a number of characteristics of rhetorical skill tied to one's "body as a whole"; accordingly, Crassus remarks, there are some unfortunates who will never achieve rhetorical excellence: those "either so tongue-tied, or so discordant in tone, or so wild and boorish in feature and gesture" that they "cannot enter the ranks of the orators" (XXV, 305). While one might argue, as Crassus does, that "strong lungs" and "vigor" are necessary for the physical demands of public speaking, his argument that "build" and "shape of the face" are also part of natural talent clearly depends on an observing, appreciative audience as a constitutive factor even in innate rhetorical ability (XXV, 305). Just as one can inquire into the conditions under which persons are granted rhetorical ability, one can also examine the circumstances under which that ability is denied: that is, the conditions that create *rhetorical disability*.[2] This understanding of rhetorical disability, which draws from the social model of disability, shifts the locus of disability from the individual (that is, a lack of rhetorical skill) to the rhetorical environment. A social model of disability encourages us to examine environmental factors such as uncut curbs, social norms of body and behavior, and ableist policies as constitutive factors in disability; a similar understanding of rhetorical disability, I would argue, directs attention to the variety of barriers that prevent certain rhetors from achieving rhetoricity with certain audiences.

Rhetorical disability is a key concern for the few rhetoricians who have examined issues of mental illness. In her essay "Rethinking Rhetoric through Mental Disabilities," for example, Cynthia Lewiecki-Wilson argues that one of the effects of the rhetorical tradition's valorization of the autonomous individual rhetor has

---

[2] For an analysis of how perceived rhetorical disability structures women's birthing plans, see Owens.

been to deny rhetoricity to persons with "severe mental impairments," who often depend on others to make their voices heard. If we are to "construct more fully humane intersubjectivity for the severely mentally disabled," Lewiecki-Wilson argues, it is necessary to expand our concept of rhetoricity to include mediated rhetorics (163). Catherine Prendergast addresses a similar issue in "On the Rhetorics of Mental Disability." A diagnosis of schizophrenia, Prendergast argues, "supplants one's position as rhetor" (47), because schizophrenic language is understood only as a "test, the record of symptoms. At best it is seen as music, as poetry, as some personal expression that has no bearing outside of itself, no transactional worth" (57). To strip one's language of the power to signify, Prendergast argues, is to remove one of the fundamental conditions of rhetoricability.

This article adds to the discussion of rhetoricability by offering a rhetorical understanding of stigma, taking the Eagleton Affair as a case study. In so doing, I hope to broaden the category of rhetorical disability to include individuals like Thomas Eagleton, who even his harshest critics would not have classified as mentally disabled in the way Prendergast and Lewiecki-Wilson describe above. Rosemarie Garland-Thomson has argued that stigma is a useful critical approach to disability because as a transitive verb, "to stigmatize" entails both a subject and an object; to take stigma as an object of criticism is to examine stigmatization as a dynamic social process rather than an individual attribute (31). As a social force enacted through language and rooted in culturally and historically contingent values, stigmatization is also a rhetorical process. As I will explore in the next section, one of the effects of that process is a public mark of bad character; accordingly, stigma also has significant implications for rhetoricability.

Stigma and the Threat of the Invisible *Kakoethos*[3]

Nearly every published discussion of stigma begins by repeating Erving Goffman's introductory note in *Stigma* about the origins of the word in the tattooing practices of ancient Greece, and this essay is no exception. Given my focus on stigma's intersections with rhetoric, however, I will tour the ancient world a little longer than most. Goffman's theory and the ancient examples both highlight how stigma serves to visibly mark the *kakoethos*.

In ancient Greece, stigma (which comes from the verb "to prick") was a rhetorical act that literally made visible those persons in the community whose invisibility was deemed a threat. Although tattooing was sometimes a decorative practice in the ancient world, as the author of the *Dissoi Logoi* tells us was customary of the Thracians (Sprague 158), it was most often used to advertise the social inferiority or moral failing of its bearer (Goffman 1). In ancient Greece, punitive tattooing was likely adopted from the Persians and continued at least until the

---

[3]I am using *kakoethos* as a theoretical, not an historical term.

ninth century, when the Emperor Theophilus punished two idolaters by tattooing "twelve lines of execrable poetry" on their faces (Jones 147). For the Greeks and the Romans (who adopted the practice as well as the name) a sentence of stigma also amounted to a demotion in social status, as it was a practice associated with slaves and foreigners (e.g., *Collected Works of Plato*, Laws 854d). In Augustan law, for example, a freed slave was usually granted the rank of full citizen; if stigmatized before manumission, however, a slave was relegated to the "lowest possible category of free non-citizens" (Gustafson 22). Even if the bearer of stigma were to have been granted the title of citizen, it is hard to imagine that he could have functioned as such; effectively, to be sentenced to stigma was to be sentenced to a life of *atimia*, without honor, a Greek punishment that stripped a citizen of many protections and rights, "primary among which was access to public space: the right to associate in the agora and speak at the assembly" (Fredal 64). The reason seems obvious: it is difficult to imagine how a bearer of stigma could be rhetorically effective when his very body announced the presence of a bad character.

Stigma was much more than a form of corporal punishment. By arresting ethical development, it was also an act of rhetorical foreclosure. As Aristotle tells us, *ethos* is not fixed: it could be cultivated by virtuous habits, as Aristotle advises in the *Ethics* (1103a17) and also constructed rhetorically by exhibiting practical wisdom, virtue, and goodwill for the audience, attributes of character that emerge "from the speech, not from a previous opinion that the speaker is a certain kind of person" (*Rhetoric* 1356a4). Stigma, however, permanently arrested one's rhetorical *ethos* at the moment of imprint, and its mark of character spoke louder, and more persuasively, than words ever could. Stigma did not subtract from a speaker's *ethos* as much as it substituted *kakoethos* in its place; stigma did not signify a lack of *ethos*, but a present *anti-ethos*.

Stigma's defining characteristic was to render its bearer's bad character permanently visible, as evidenced by the placement of the marks on the face and hands, the two areas of the body least able to be hidden from public view (Gustafson 21). This visibility served a civic function in addition to a punitive one, as the stigmatized body functioned as a monument to the law.[4] As Protagoras explained to Socrates in the eponymous dialogue, the objective of punishment in general is "to prevent either the same man or, by the *spectacle* of his punishment, someone else, from doing wrong again" (324b, my emphasis). As a living, breathing, inscription of the law and a spectacular example of the consequences of its

---

[4] The visibility of the mark also operated as a kind of risk management. If a man could not be trusted to learn from punishment, he at least could be made visible so that others knew to guard themselves against him, which was facilitated by the fact that punitive stigma often indicated its bearer's specific crime. A man imprinted as "thief" was likely to be watched carefully when entering the market, and one knew not to believe the words of a Roman who wore the letter "K" for *kalumnia* (bearing false witness) in the middle of his forehead. Separated from the crowd and aware of his permanent visibility, the bearer of stigma would police himself: as a form of discipline and punishment, stigma was a remarkably efficient form of social control.

transgression, stigma inculcated the power of *nomos* in the community. In some respects, we might read stigma as an embodied form of epideictic, for it put its bearer on permanent display as a blameworthy exemplar of *kakia*. Whether by another's speech or the sharp prick of stigma, the bad character is revealed by epideictic discourse because his unacknowledged, invisible presence poses a threat to community values, and by extension, the community itself. If epideictic praise displays the exemplar of *arete* in order to encourage "mimetic action" and propagate public virtue (Hauser 15), epideictic blame displays the exemplar of *kakia* in order to discourage mimesis and "reinstall" the *nomos* through an act of public renunciation, to borrow from Judith Butler's reading of the consequences of publicly breaking taboo (116).

The Demand for Disclosure

In contemporary usage, "stigma" now generally refers to an individual's disgraced social identity instead of its bodily mark (Goffman 2). Writing in 1963, Goffman's examples of stigma included bodily "abominations" (such as amputation or severe scarring), "blemishes of character" (such as illiteracy, prostitution, or criminal conviction), and "tribal" stigmas of race, nation, and religion (4). Although it is jarring, to say the least, to place a woman with a mastectomy and a "castrated Norwegian male sex offender" (75) in the same category, for Goffman the defining element of stigma is that its bearer is seen as "deeply discredit[ed]" in a particular culture at a particular time (3).[5] In each case,

> an individual who might have been received easily in ordinary social intercourse possesses a trait that can obtrude itself upon attention and turn those of us whom he meets away from him, breaking the claim that his other attributes have on us.... By definition, of course, we believe the person with a stigma is not quite human. On this assumption we exercise varieties of discrimination, through which we effectively, if often unthinkingly, reduce his life chances. (5)

While contemporary stigmas may not be literally imprinted on the body, visibility plays a crucial role in Goffman's theory of stigma. Visibility, as I will be using it in the remainder of this article, not only designates literal visibility, but also epistemological visibility—to be visible as stigmatized is to be known as a person of "spoiled identity": "tainted," "discounted," of a "less desirable kind," and "in the extreme, a person who is quite thoroughly bad, or dangerous, or weak" (3). Stigma is more than just a single attribute or action deemed undesirable according to a particular community's norms and values: it entails the "global devaluation"

---

[5]Thanks to Michael Bérubé for drawing my attention to this passage. Bérubé wonders if these "weird" portions of the text might be where "Goffman is winking at us, as one of the 'wise': he knows that stigma has a temporal dimension, that social opprobrium, like everything else, can be historicized" ("Special" n.p.).

of a person's character on the basis of that attribute (Katz, qtd. in Brouwer 117). A visible stigma operates as a master hermeneutic that guides the interpretation of all other facets of character, as one of Goffman's interviewees, a criminal, observes:

> I always feel . . . with straight people—that whenever they're being nice to me, pleasant to me, all the time really, underneath they're only assessing me as a criminal and nothing else. It's too late for me to be any different now to what I am, but I still feel this keenly, *that that's their only approach,* and they're quite incapable of accepting me as anything else. (15, my emphasis)

One of stigma's most insidious aspects, then, is to designate its bearer as an object of information that is frustratingly general—to be stigmatized is to be known, simply, as *bad.* That generality, I would argue, allows slippage between the many permutations of badness: worthless, evil, dirty, ugly, weak, cowardly, envious, dangerous—to choose just a few of the many variants of *kakos* in ancient Greek.[6] This slippage is especially evident in the stigma of disability, which has a history of being read as a sign of evil and associated with weakness, criminality, asexuality, vagrancy, dangerousness, and worthlessness (cf. Garland-Thomson 36–37).

Goffman's note that stigma "break[s] the claim" of other attributes may be attributed to the power of stigma's *kakoethos.* If one's attributes make a claim, it is a claim of character—a stigmatized attribute breaks that claim by offering a more persuasive argument for a particular audience, and much like the ancient Greek mark, it drowns out other forms of "speech" about character. One might be unfailingly kind, breathtakingly beautiful, and a whiz at calculus, but if one walks with a cane, wears the hijab, or is known to have bipolar disorder, these attributes tend to shout down the others in rhetorical environments where cane use, the hijab, or bipolar disorder are stigmatized.

For those whose stigmas are not immediately visible, those Goffman calls the "discreditable," it becomes necessary to rhetorically manage information about the stigmatized attribute so as to preserve one's *ethos* and prevent the slide from potentially discreditable to actually discredited: "to display or not to display; to tell or not to tell; to let on or not to let on; to lie or not to lie; and in each case, to whom, how, when, and where" (Goffman 42). To live as discreditable is to pass for "normal," which sometimes (though not always) entails "a very high level of anxiety, in living a life that can be collapsed at any moment" (87). In part, anxiety over passing stems from the pressures of living a secret life which, as Edwin Black explains, is categorically different from a private life:

> A private life is simply one shared only with intimates, conducted without attracting notice. There is an expectation that everyone will have a private life.

---

[6]For more on the many meanings of *kakos* in ancient Greek texts, see the excellent essays in Sluiter and Rosen, especially Mulhern, who discusses its meaning in Aristotle.

> But a secret life is one attended by potential scandal, one in which there is a disparity between appearance and reality, between reputation and character. To say that someone has a private life is no more than to report a common expectation, but to say that someone has a secret life is to imply something unsavory. (145)

To live a secret life is not just to live a life where information is concealed; importantly, it is to live with the understanding that one's secret *ought* to be disclosed—the "potential scandal" of the secret life implies that it is a secret that ought not be kept, and a secret that ought not be kept, writes Black, is considered "public property" (141).

To suppress information about one's stigmatized attribute, that is, to pass for "normal," is to withold information that is not perceived to belong to you. The invisible *kakoethos*, because she violates the *nomos* in some way, either through an unruly body, unsanctioned behavior, or devalued identity, poses a threat to the community and as such, has a duty make herself visible to them. The dilemma of the discreditable, the invisible *kakoethos*, is a familiar one: out yourself, or risk being outed and face additional consequences. Because passing entails concealing the discrepancy between the "false" character (*ethos*) and the "real" character (*kakoethos*), to pass is to enjoy benefits that would be lost were one to be revealed, discredited, and declared a fraud.

In the decade before Thomas Eagleton made headlines as the "running mate who wasn't" (Wilkinson), it is clear that he strategically managed information about his mental illness in order to safeguard his invisibility and protect his *ethos*. On 17 December 1960, shortly after winning the election for attorney general of Missouri, the *St. Louis Post Dispatch* reported that Eagleton had been admitted to Barnes Hospital in St. Louis "suffering from a virus." His illness, the story continued, "was complicated by hard work in his successful campaign." Four years later, after he had been elected lieutenant governor, the *Dispatch* reported that Eagleton had lost nearly fourteen pounds in two weeks and had entered the Mayo clinic for "tests." Two years after that, the *Dispatch* again reported that Eagleton had been admitted to Johns Hopkins University hospital, this time with a "gastric disturbance." He again underwent "tests" (Kifner). Although Eagleton was later to claim that he had not deceived George McGovern, there's no question that the information given to the *St. Louis Post Dispatch* was not entirely accurate. In 1960, Eagleton had indeed been admitted to Barnes Hospital, but not for a virus: he stayed in the Renard Psychiatric Division for four weeks and received talk therapy, medication, and ECT to treat nervous exhaustion and depression. He was hospitalized at the Mayo Clinic, not Johns Hopkins, in 1966, where he received a similar course of treatment (McGovern 202).

Based on what was disclosed to the *Dispatch* and what was concealed, it is readily apparent where the lines of stigma are drawn. It is acceptable to be physically ill, acceptable to be hospitalized, acceptable to be suffering from a virus, and even acceptable to be exhausted—after all, to be hospitalized for "hard work" seems like the sign of a strong character, not a weak one—but to be hospitalized for depression, and particularly to have received physical treatment for that depression, was discrediting information. As a stigmatized attribute, Eagleton's depression was subject to the demands of visibility, which motivated the tandem desires for secrecy and disclosure: the secret is a resistance to the demands of disclosure, an intentional refusal to make oneself visible. Thomas Eagleton's mental illness could never have been merely a private matter. Inasmuch as it was a stigmatized attribute and subject to the demands of visibility, it could only ever have been a secret affair, and as the events of 1972 vividly illustrated, its discovery the subject of scandal.

Exhibit A: Thomas Eagleton

> Tom Eagleton is going to be one of the most visible vice-presidential candidates in history. If that were considered desirable before his press conference Tuesday, it is considered absolutely essential now. He will be Exhibit A of Tom Eagleton's soundness and stability, and he will be campaigning in the full knowledge that his audiences will note every tic, twitch, or quaver for signs that the doctors fell short of their goal. (Sackett)

In many ways, this statement, taken from an editorial that ran immediately after Eagleton's public announcement, sums up how stigma functioned in the Eagleton Affair, speaking both to the demand ("absolutely essential") for visibility as well as the threat of that visibility to his rhetoricability. Before we get to those aspects of the Eagleton Affair, however, it is necessary to explain what the Affair was *not*: although it might seem counterintutive, the Eagleton Affair was not about Thomas Eagleton's ability to serve as vice president of the United States. I make this brief detour in order to counter the possible objection that Eagleton was dropped from the ticket because he was unfit or unqualified to discharge the duties of the office. Eagleton was not forced to resign from the McGovern campaign because he was impaired; he was forced to resign because his presence was perceived to "cripple" the campaign (Salinger), an issue of *ethos* rather than competence or qualifications.

There is no constitutional language or other federal regulation that sets a standard for physical or mental health for the office of president or vice president of the United States. The only language that speaks to the requirements of the presidency is section I, article 2 of the constitution, which outlines the citizenship, residency, and minimum (not maximum) age restrictions for eligible candidates. Because there are no official measures in place, the campaign functions as a screening mechanism to determine a candidate's health (Robins and Post). In the absence of official regulations, in other words, a political candidate's physical or mental fitness is diagnosed in the clinic of public opinion. The practice of "vetting" a political candidate is of relatively recent vintage: McGovern comments that with the exception of Johnson's review of Hubert Humphrey in 1964, he was unaware of any precedent in which the background of a potential vice president had been subject to a vetting process (200). The term "vetting" comes from the practice of evaluating a horse's fitness to compete in an upcoming race, and the practice of political vetting operates in much the same way: it does not determine a candidate's ability to serve in office, it determines a candidate's ability to *run* for it by culling, to return to Aristotle's definition of rhetorical *ethos*, those who fail to exhibit practical wisdom, virtue, or goodwill. Vetting is a practice of making visible, of rooting out the invisible *kakoethos* who may rhetorically disable the campaign.

There is a difference between raising questions about a candidate's health and judging that a particular condition categorically disqualifies a candidate because of perceived unelectability, and that difference, I contend, is found in stigma's *kakoethos*. Of the past twenty presidents, fourteen have suffered "significant" illness during their terms (Abrams 115). Although in many cases these illnesses were concealed or their severity was downplayed (e.g., Roosevelt, Kennedy, Reagan), candidates for president and vice president have run successful campaigns even after serious health concerns were disclosed to the American public. Eisenhower, for example, was elected to a second term in 1956 after suffering a major heart attack in office, and Dick Cheney's heart condition was a perennial subject of late night comedy, but not grounds to keep him off the ticket in 2000 and 2004.

The stigma of mental illness, however, and particularly the stigma of psychiatric treatment "is so deep and pervasive in the public mind that the fact that such treatment was needed is seen as a *sign of weakness* and *instability* sufficient to disqualify the candidate" (Robins and Post 852, my emphasis). Unlike heart disease (Cheney) or melanoma (McCain), mental illness is considered by many not just to be the sign of a bad character, but *caused* by bad character. A contemporaneous description of the Eagleton Affair, for example, explained that in 1972, mental illness "still carried a stigma, like venereal disease" (White 209), another condition often attributed to bad character. Although attitudes about mental illness have shifted since the time of the Eagleton Affair, a study as recent as 2006 found that thirty-three percent of respondents were willing to attribute the cause of major depressive disorder to a person's "own bad character." The same percentage believed that a person with depression was "dangerous" and "likely to hurt others" (Schnittker 1374–1375). It's no wonder, then, that the prospect of mental illness in the Commander in Chief "invokes a feeling of dread in the public" (Robins and Post 852).

A diagnosis of mental illness, no matter how far in the past, no matter the severity of the condition, no matter the circumstances, is a permanent identity. One "has," or "suffers from" heart disease. One is perceived "to be" mentally ill *even after successful treatment*. In much the same way, Goffman writes, even if one is able to "correct" a stigmatized attribute, identity is transformed not from stigmatized to "normal," but "from someone with a particular blemish into someone with a record of having corrected a particular blemish" (9). After diagnosis, Goffman argues, one is always a mental patient in the same way that one is always a convict: an ex-mental patient, an ex-con. There is always a record to be concealed, there is always the potential for discovery, and there is always the weighty knowledge that this information *ought* to be disclosed. "When you're in politics," Eagleton remarked, "you're in some kind of fishbowl, and thus it always goes through your mind: When will I be asked? Or when will somebody know?" (Lydon).

That moment came shortly after Eagleton's nomination, when rumors about his mental health began to filter in to the McGovern campaign team. At first, McGovern recalls, the information about Eagleton's medical history "seemed a minor matter of little consequence" (201). Eagleton was to appear on *Face the Nation* the following week; in an echo of the original *Dispatch* reports, McGovern's campaign manager Frank Mankiewicz advised Eagleton to explain "he was such a dedicated campaigner that he had campaigned himself right into the hospital" if the subject of the hospitalizations was raised (McGovern 201). On 17 July, however, an anonymous caller contacted a worker in McGovern's Washington head-quarters to report that Eagleton had been hospitalized on three occasions not for exhaustion, but for depression, and that he had received electroconvulsive therapy during two of his hospital stays (McGovern 202). The caller also said he had already relayed this information to reporter John Knight at the *Detroit Free Press*, who was told that Eagleton had been hospitalized for "a manic-depressive state with suicidal tendencies" (Giglio 656).[7] Aware that the story was about to break, McGovern writes, the campaign was faced with an "immediate decision" to pub-licly disclose the information about Eagleton's psychiatric history (204), which they hoped would put the matter to rest.

At a hastily organized press conference with George McGovern on 25 July, Eagleton revealed to reporters the details of his hospitalizations, including the fact that he had received electroconvulsive therapy. Although Eagleton did use the word "depression," his focus, following Mankiewicz's recommendation, was on exhaustion: "as a young man, I drove myself too hard," he explained, and described his hospitalizations as the result of "overexertion in politics." He would avoid such fatigue in the future, he noted, by "pacing himself" and keeping firm to his practice of not working on Sundays (Lydon). By directing his audience's attention to the actions that led to the hospitalization, Eagleton sought to rhetori-cally manage stigma by describing his illness in physical, rather than mental, terms.

---

[7]Giglio claims that Eagleton's actual diagnosis was bipolar II disorder, information he found in Eagleton's FBI file and confirmed with the psychiatrist who treated Eagleton in the eighties (659). Because the bound-aries of psychiatric nosology are notoriously blurry, I am wary of claiming that Eagleton was "really" bipolar, and have opted to describe his illness as depression in this article.

This is best illustrated by his description of the 1964 hospitalization, which was to treat "stomach problems": "I'm not too different from the fellow in the Alka-Selzer ad who says 'I can't believe I ate the whole thing' . . . I do get a nervous stomach situation." Eagleton said he was currently in fine health, although he joked he was "two pounds overweight and [had] half a hemorrhoid" (Lydon). Despite Eagleton's attempts to reterritorialize the issue from mind to body, most papers went with a story nearly identical to the front page story from the *New York Times*, which ran under the headline "Eagleton Tells of Shock Therapy on Two Occasions." Positioned next to the story is a large picture of a smiling McGovern and a pensive Eagleton, who holds his hand to his forehead. This image, holding one's head in hand, is the quintessential visual representation of depression. Notably, this image is also a visual depiction of shame.

After the press conference, although the opinions on whether Eagleton should be removed from the ticket varied wildly, the editorial pages were "almost unanimous" in their criticism of Eagleton's failure to reveal his medical history to McGovern at the moment he was asked to join the ticket (Kreger 30). It was not only McGovern who had a right to know, one editorial in the *Chicago Tribune* intoned, the American voters had a right to know as well. Although the editorial maintained that "the fact that a man has had psychiatric counseling and even shock treatments should not in itself disqualify him forever for responsible public service," it insisted that Eagleton was obligated to reveal his "skeleton" to the American public. "Mr. Eagleton may still be able to erase the blot on his medical record," the editorial concluded, but only "if he recognizes that the people are entitled to know all of the facts about the emotional stability of a man who could become President" ("Sen. Eagleton's Past").

The effect of Eagleton's disclosure on his rhetoricability was considerable. Eagleton was burdened with the stamp of bad character, which was compounded by the accusations of fraud and deception that accompanied his forced outing; in addition, to be visible as mentally ill also carried a threat to rhetoricability that I will call the "diagnostic hermeneutic," an interpretive frame in which the audience takes on the diagnostic gaze of the physician, searching body and speech for symptoms. The roots of this hermeneutic are similar to what Prendergast describes as the denial of signification to schizophrenic speech and writing, in that mental illness is almost always interpreted via language. It would be inaccurate to describe Thomas Eagleton's speech as evacuated of signification, however. If anything, Eagleton's speech signified too much, illustrated by the editorial I used as the epigraph for this section, which warned that "his audiences will note every tic, twitch, or quaver for signs that the doctors fell short of their goal." Read through the diagnostic hermeneutic, even the smallest bodily movements—this gesture, that facial expression, this stammer—act as signs, symptoms of the illness hidden beneath, and in Eagleton's case, taken as a warning that someone of unsound mind might be about to assume a position that demanded extraordinary reason and judgment. For most people, trembling hands might be the sign of too much caffeine. Within the diagnostic hermeneutic, however, a tremble is never just a tremble. One can only imagine the campaign literature the Nixon team would have crafted on the theme of Eagleton's "trembling finger on the nuclear button" (Hart, qtd. in Giglio 671).

Media reports of the press conference invariably read Eagleton's physical comportment through the diagnostic hermeneutic. The *New York Times*, for example, described the senator as "manifestly nervous," revealed by the fact that his "hands and his face" seemed to "quiver slightly" (Lydon). During Eagleton's *Face the Nation* appearance at the height of the Affair, the moderator wondered out loud during the program if "there was some unsettling significance to Eagleton's perspiration and the tremors in his hands" (McGinniss 30). After the show aired, one of Eagleton's colleagues in the Senate complimented him for keeping his temper when interacting with a reporter who had falsely accused him of driving drunk. "Well, I had to," Eagleton replied, "if I had gotten mad everybody would have said, look at that Eagleton, no self-control, he doesn't belong on the ticket" (McGinniss 30).

The Contingency of Stigma

How we understand the Eagleton Affair as historically or politically salient depends on the frame used to tell the story, and these narrative frames are usually organized according to competing rhetorics of blame in which Eagleton and McGovern take turns as victim and perpetrator. One version of the story characterizes the Eagleton Affair as Eagleton's failure to disclose his psychiatric history. On the final night of the Democratic National Convention, when McGovern's campaign director Frank Mankievicz asked if there were any "skeletons rattling around in [his] closet" (Giglio 653), Eagleton replied in the negative, withholding information he knew to be damaging to his *ethos* from the McGovern campaign in order to further his political ambitions. A second version of the story blames McGovern and his campaign team. In a rush to choose a running mate, the McGovern campaign settled on Eagleton without vetting him. This story assumes that a thorough background investigation would have uncovered Eagleton's mental illness and hospitalizations, which would have disqualified him from consideration. Eagleton also held McGovern responsible for the events that transpired in late summer 1972, but for another reason. For Eagleton, who claimed to have "tucked away the memory of mental illness as completely as the memory of a broken leg," information about his mental health or prior hospitalization was simply not relevant to Mankievicz's question about his past (White 209). In this version of the Affair, neither mental illness nor its treatment were "skeletons," and by making them so, the McGovern campaign broadcast the message that seeking treatment for mental illness was something to be ashamed of, disqualifying a substantial portion of the American populace from the second-highest political office in the country. Although McGovern and Eagleton are the most frequent subjects of blame, it is clear that stigma, which motivated Eagleton's secrecy as well as the McGovern's demand for disclosure, was the rhetorical engine of the Eagleton Affair.

Stigma's contingency, however, makes it a slippery target for blame. Shortly after the Eagleton Affair had ended, George McGovern attended a Washington Redskins exhibition game where, he writes in his memoirs, he was greeted by boos and jeers. At about the same time, Thomas Eagleton was introduced during a St. Louis Cardinals game, where he was treated to a "rousing standing ovation" from his fellow Missourians (McGovern 192)—a Lacedemonian praised in Lacedemonia, as Aristotle might observe. From McGovern's perspective, those eighteen days in 1972 "strengthened" Eagleton, "at least with his Missouri constituents. They saw him as a gallant martyr, the victim of a heavy-handed, callous and conniving McGovern campaign. That was surely the view, at the time, of many other Americans" (191). After he was dropped from the 1972 ticket, Eagleton was the subject of supportive stories from *Life* and *Newsweek,* and he is represented as a sympathetic character in Theodore White's definitive account of the election, *The Making of the President.*[8] He continued to serve in the U.S. Senate until 1987, having never lost an election. Was it the case that the stigma of mental illness was not as strong as both McGovern and Eagleton believed it was? Or was it the case that the flurry of rhetorical activity of the Eagleton Affair allowed Eagleton to develop *ethos* not in spite of, but because of his visible difference? As Eagleton himself put it to George McGovern in a last-ditch effort to stay on the ticket: "George, I am no longer Tom Who. I am Tom Eagleton, suddenly a very well-known political figure" (White 217).

Here is where contemporary stigma diverges considerably from its ancient mark. Contemporary stigma might be a powerful social and rhetorical force, but its mark is not indelible—the contemporary mark of stigma can be erased as social norms and values shift and change. *Kakoethos* is only present when the audience recognizes it as such. In an ironic turn, many believe that it was not Eagleton who cost McGovern the election, but McGovern's decision to drop him, an action that cost McGovern his *ethos* as a "politician who would not, like others, do *anything* to get elected" (218).

Despite his distinguished career in the Senate, which included authoring the Eagleton Amendment (which cut off federal funding for the bombing of Cambodia and effectively ended the Vietnam War), and pivotal roles in the Clean Air and Water Acts, Thomas Eagleton is still best known for those eighteen days in 1972. Eagleton was acutely aware of his legacy: he wryly commented in a 2003 interview with the Associated Press that "in my obituary, it will probably be, 'Tom Eagleton, United States senator for Missouri, for a short time the vice presidential candidate

---

[8]Eagleton believed White's book to be the "best" account of the 1972 election (Eagleton).

on the McGovern ticket in 1972,' so that will be in my first paragraph." When Eagleton died in 2007, he was proven right (Clymer).

Fighting the Stigma of Mental Illness

In December 2009, former Minnesota senator and current gubernatorial candidate Mark Dayton announced that he had "long" been medicated for depression. Unlike Eagleton, there was no breaking news story that prompted the disclosure; well aware of the visibility demands of stigma, however, Dayton attributed his decision as a response to the public's "right to know" (Sturdevant).[9] Because he has not made his treatment regimen public, one political commentator observed that "it looks less like he's trying to protect his privacy and more like he is trying to hide something" and advised Dayton to provide "additional clarification" or it would haunt the rest of his campaign (Stassen-Berger and Helgeson). Although many aspects of Dayton's disclosure and subsequent press coverage are reminiscent of the events of 1972, there is a remarkable difference: Dayton clearly believed that disclosing information about his depression would not disqualify him from the race. That Dayton released information about his psychiatric history in response to a perceived public demand suggests that the stigma attached to depression is still quite powerful; however, Dayton's willingness to speak about his illness suggests that the stigma of depression has begun to subside since Eagleton's time.[10] Dayton is not the only politician in recent years to reveal a history of mental illness: Lawton Chiles was elected twice as Florida's governor after revealing that he had been treated for depression, and Rhode Island voted Patrick Kennedy into the U.S. House of Representatives even after Kennedy disclosed that he was undergoing treatment for bipolar disorder.

---

[9]To voluntarily make oneself visibly stigmatized in this way, as Dan Brouwer argues in his analysis of HIV/AIDS tattoos, is a "precarious" form of rhetorical action that entails a significant amount of risk. On the one hand, Brouwer argues, the self-stigmatizer is motivated by and participates in a long tradition of visibility politics, which "assume that 'being seen' and 'being heard' are beneficial and often crucial for individuals or a group to gain greater social, political, cultural or economic legitimacy, power, authority, or access to resources" (118). On the other hand, to voluntarily make oneself visible is to "summon surveillance and the law" (Brouwer 116, quoting Phelan 6). For a critique of the demands for visibility in the disability and LGBT communities, see Samuels.

[10]Although I do not wish to offer a *post hoc* explanation for this shift, I believe that it is related to growing characterization of depression as the common cold of mental illness. An argument might even be made that depression has ceased to be regarded as a mental illness in the public imagination; although a person with schizophrenia is readily identified as mentally ill, a person with depression is not. Although belief in the disease model, supported by the hyper-popularity of selective serotonin reuptake inhibitors (SSRIs), may contribute to the decrease in depression's stigma, I believe the glut of literature about depression in the last two decades has played an equal, if not stronger, force in shaping the public perception of depression. While I'm not naïve enough to claim that memoir alone has the power to abolish stigma, it lays the foundation for rhetoricability, which is an effective tool in that fight. Katie Rose Guest Pryal's article in this issue explores how such a process works in one genre of memoir.

In recent years, federal agencies like the National Institutes for Mental Health (NIMH) and advocacy groups like the National Alliance on Mental Illness (NAMI) have identified stigma as one of their top priorities. A major focus of anti-stigma campaigns is to educate the American public about the biological nature of mental illness, a strategy we might think of as the "medical literacy" approach, which promotes the idea that mental illness is a disease "like any other" (Read et al.). The goal of the disease model is to shift cause (and implicitly, blame) from the self or the environment to genes and chemistry: because mental illness is biological, the argument goes, it is "not anyone's fault and therefore should have no stigma" (Dumit 41).

One such literacy program touted as a "powerful anti-stigma tool" is the "In Our Own Voice" (IOOV) program sponsored by National Alliance on Mental Illness (NAMI). IOOV offers presentations by people living with major depression, bipolar disorder, and schizophrenia in order to change their audience's "hearts, minds, and attitudes" about mental illness (NAMI). As of 2007, IOOV has reached over two hundred thousand people with its presentations, which are offered to community groups free of charge. A 2006 study argues that IOOV has been effective at reducing stigma in its target audiences (Wood and Wahl 50). However, it is important to note how stigma is defined in this study. Stigma, the authors contend, "frequently stems from a lack of accurate knowledge about mental illnesses, as mental illness in general is a topic that is frequently not discussed and oftentimes not openly addressed. Limited opportunities for education and learning, then, allow the misunderstanding of mental illness to continue" (46). If stigma results from a lack of correct medical information, one combats stigma by increasing medical literacy. Accordingly, the IOOV program emphasizes the disease model and focuses on the varieties of treatment and "coping strategies" that led its speakers from "dark days" to "recovery and hope" (NAMI). It is worth noting that IOOV was made possible via a grant from the pharmaceutical company Eli Lilly and its participants are referred to as mental health "consumers" throughout the literature.

As a means to fight stigma, the strategy of medical literacy has the potential to backfire. Some studies have found, for example, that emphasizing the disease model leads to the perception that people with mental illness are different in kind, categorically distinct from "normal" people (Mehta and Farina; Read et al.; Schnittker). Attributing cause to genetics may produce an even worse result:

> Genetic arguments may work in an asymmetric fashion—they encourage the view that mental illness is impersonal and uncontrollable in its development, but more stable and unyielding in its course. By the same token, genetic arguments inflate perceptions of dangerousness insofar as the mentally ill are always at risk for violence, even when treated. The 'depth' implied by genes entails a latent threat. (Schnittker 1372)

The problem with the medical literacy campaigns, as I see it, is not the disease model *per se*. The problem with these campaigns is that they misunderstand how stigma operates. If stigma is a matter of values rather than facts, whether mental illness has its origins in genes, chemistry, biography, environment, bad character, God's will, or the cycles of the moon is of little importance. If stigma is not passive ignorance but the active rhetorical propagation of community norms and values coupled with the demand for visibility, anti-stigma campaigns that focus on medical literacy miss their target and run the risk of subverting their own goals.

One might point out, and rightly so, that the participants in campaigns like IOOV are clearly rhetorically effective. To be rhetorically disabled does not mean rhetorically *un*able. Lewiecki-Wilson observes, for example, that a number of mentally ill writers have "moved audiences and brought about change in the treatment of the mentally ill" (165). This is true, and it brings me to a final point about the relationship between stigma, *ethos*, and rhetorical disability. *Ethos* is contingent not only upon rhetor, audience, and social values, but also on subject matter. Each of the rhetors Lewiecki-Wilson identifies is rhetorically effective when speaking or writing on the subject of mental illness, as are the rhetors of the IOOV program. In these cases, to have borne the stigma of mental illness serves to support a rhetor's *ethos*, not damage it—it is evidence of practical wisdom about the subject at hand. But, in many cases, this *ethos* is limited to *only* the subject at hand. Part of stigma's disabling effect is to restrict a stigmatized rhetor to the position of spokesperson, to deny what we might think of as ethical flexibility—the capacity to speak with authority on a number of different topics—a rhetorical ability we demand from our political leaders. Maybe we will finally have moved beyond the Eagleton Affair *not* at the moment when a presidential candidate can disclose a history of depression and still be elected to national office, but when a presidential candidate can discuss (not "disclose") a history of depression during a press conference and receive a follow-up question about her feelings on the economy.

Conclusion: Turning Toward *Kakos*

Tobin Siebers has argued that postmodern studies of the body have often focused on pleasure at the expense of pain; the body of body theory is usually "a body that feels good and looks good—a body on the brink of discovering new kinds of pleasure, new uses for itself, and more and more power" (Siebers 742; see also Davis 5). In much the same way, rhetorical studies has devoted itself for millennia to propagating the "good man speaking well," and producing rapturous encomia on rhetorical achievement.[11] We turn toward people who look, think, act, and

---

[11]On this point, it is telling that despite rhetoricians' love of epideictic, *psogos* speeches, speeches of blame, are almost entirely absent from the rhetorical record. See Rountree, who provides one of the few sustained discussions of *psogos* in the literature.

speak according to our criteria for the beautiful, the healthy, the good, the talented, and the eloquent, and we turn away from those who violate these powerful edicts of the *nomos*. When rhetorical disability enters the attention of rhetorical scholars, it is usually as a moralistic tale of overcoming: consider that one of our field's founding myths is the story of Demosthenes, the rhetorical supercrip who corrects a stutter to become the ancient world's most celebrated orator. We explore, often with our students, how writers and orators might best accommodate the audience, but rarely examine the limits of that accommodation or question the idea that the responsibility of accommodation falls on the rhetor. What might a "reasonable" rhetorical accommodation look like, to borrow from the language of the Americans with Disabilities Act? What might happen if we directed critical attention to the rhetorical environment in which a tied tongue, trembling hands, an "unshapely" face or body, mental illness, or other contributors to *kakoethos* threaten rhetoricability? What might we do with the bad man speaking poorly?

References

Abrams, Herbert L. "Can the Twenty-Fifth Amendment Deal With a Disabled President? Preventing Future White House Cover-Ups." *Presidential Studies Quarterly* 29.1 (1999): 115–133. *EBSCO*. PDF file. 10 March 2010.

Aristotle. *Nicomachean Ethics*. Trans. Terence Irwin. Indianapolis, IN: Hackett Publishing, 1999. Print.

———. *On Rhetoric: A Theory of Civil Discourse*. Trans. George A. Kennedy. New York/Oxford: Oxford University Press, 1991. Print.

Bérubé, Michael. "Special 'Special' Edition." *American Airspace* 2 April 2009. Web. 10 April 2010. <http://www.michaelberube.com/index.php/weblog/special_special_edition/>

Black, Edwin. "Secrecy and Disclosure as Rhetorical Forms." *Quarterly Journal of Speech* 74.2 (1988): 133–150. *EBSCO*. PDF file. 22 February 2010.

Brouwer, Dan. "The Precarious Visibility Politics of Self-Stigmatization: The Case of HIV/AIDS Tattoos." *Text and Performance Quarterly* 18.2 (1998): 114–136. *JSTOR*. PDF file. 15 March 2010.

Brueggemann, Brenda. *Lend Me Your Ear: Rhetorical Constructions of Deafness*. Washington, DC: Gallaudet University Press, 1999. Print.

Butler, Judith. *Excitable Speech: A Politics of the Performative*. New York: Routledge, 1997. Print.

Cicero. "De Oratore." *The Rhetorical Tradition: Readings from Classical Times to the Present*, 2nd Ed. Patricia Bizzell and Bruce Herzberg. New York and Boston: Bedford/St. Martin's, 2001. Print.

Clymer, Adam. "Thomas F. Eagleton, 77, a Running Mate for 18 Days, Dies." *New York Times* 5 Mar. 2007: n. pag. Web. 5 March 2010.

Davis, Lennard J. *Enforcing Normalcy: Disability, Deafness, and the Body*. London: Verso, 1995. Print.

Dumit, Joseph. "Is It Me or My Brain? Depression and Neuroscientific Facts." *Journal of Medical Humanities* 24.1/2 (2003): 35–47. Print.

Eagleton, Thomas. E-mail to the author. 6 October 2006.

"Eagleton's Own Odyssey." *Time* 7 August 1972. Web. 2 March 2010.

Fredal, James. *Rhetorical Action in Ancient Athens: Persuasive Artistry from Solon to Demosthenes.* Carbondale: Southern Illinois University Press, 2006. Print.

Garland-Thomson, Rosemarie. *Extraordinary Bodies: Figuring Physical Disability in American Culture and Literature.* New York: Columbia University Press, 1997. Print.

Giglio, James N. "The Eagleton Affair: Thomas Eagleton, George McGovern, and the 1972 Vice Presidential Nomination." *Presidential Studies Quarterly* 39.4 (2009): 647–676. EBSCO. PDF file. 18 February 2010.

Gizbert, Richard. "Sarah Palin: Thomas Eagleton, the Sequel." *Huffington Post* 1 September 2008. Web. 17 February 2010.

Goffman, Erving. *Stigma: Notes on the Management of Spoiled Identity.* New York: Simon & Schuster, 1963. Print.

Green, Joshua. "The Eagleton Scenario." *The Atlantic* 2 September 2008: n. pag. Web. 17 February 2010.

Gustafson, Mark. "The Tattoo in the Later Roman Empire and Beyond." *Written on the Body: The Tattoo in European and American History.* Ed. Jane Caplan. Princeton: Princeton University Press, 2000. 17–31. Print.

Ham, John. "Is Biden Another Eagleton?" *John Locke Foundation* 29 August 2008. Web. 5 March 2010.

Hamilton, Edith, and Huntington Cairns. *The Collected Works of Plato.* Princeton: Princeton University Press, 1989. Print.

Hauser, Gerard A. "Aristotle On Epideictic: The Formation of Public Morality." *Rhetoric Society Quarterly* 29.1 (1999): 5–23. JSTOR. PDF file. 12 February 2010.

Jones, C. P. "Stigma: Tattooing and Branding in Graeco-Roman Antiquity." *The Journal of Roman Studies* 77 (1987): 139–155. JSTOR. PDF file. 17 February 2010.

Katz, Irwin. *Stigma: A Social Psychological Analysis.* New York: Lawrence Erlbaum, 1981. Print.

Kifner, John. "The Rise and Fall of Tom Eagleton." *New York Times* 1 August 1972: n. pag. *ProQuest Historical Newspapers.* PDF file. 10 December 2007.

Kreger, Donald S. "Press Opinion in the Eagleton Affair." *Journalism Monographs* 35.1 (1974): 1–44. ERIC. PDF file. 20 February 2010.

Lewiecki-Wilson, Cynthia. "Rethinking Rhetoric through Mental Disabilities." *Rhetoric Review* 22.2 (2003): 156–167. Print.

Lydon, Christopher. "Eagleton Tells of Shock Therapy on Two Occasions." *New York Times* 26 July 1972: n. pag. *ProQuest Historical Newspapers.* PDF file. 10 December 2007.

McCain, Robert Stacy. "Obama's Eagleton Affair." *The American Spectator* 29 April 2008: n. pag. Web. 5 March 2010.

McGinniss, Joe. "Out of the Spotlight, Eagleton Reassembles His Image." *Life* 18 August 1972: 30–31. Print.

McGovern, George. *The Autobiography of George McGovern: Grassroots.* New York: Random House, 1977. Print.

Mehta, Sheila, and Amerigo Farina. "Is Being Sick Really Better? Effect of the Disease View of Mental Disorder on Stigma." *Journal of Social and Clinical Psychology* 16.4 (1997): 405–419. Print.

Mulhern, J. J. "*Kakia* in Aristotle." *Kakos: Badness and Anti-Value in Classical Antiquity.* Ed. Ineke Sluiter and Ralph M. Rosen. Mnemosyne: Supplements. History and Archaeology of Classical Antiquity Vol. 307. Leiden/Boston: Brill, 2008. Print.

National Alliance on Mental Illness. "What is IOOV?" N.d. 10 April 2010. http://www.nami.org/template.cfm?section=In_Our_Own_Voice

Owens, Kim Hensley. "Confronting Rhetorical Disability: A Critical Analysis of Women's Birth Plans." *Written Communication* 26.3 (2009): 247–272. *Sage Journals Online.* PDF file. 20 May 2010.

Phelan, Peggy. *Unmarked: The Politics of Performance.* New York: Routledge, 1993. Print.

Prendergast, Catherine. "On the Rhetorics of Mental Disability." *Embodied Rhetorics: Disability In Language and Culture.* Ed. James C. Wilson and Cynthia Lewiecki-Wilson. Carbondale: Southern Illinois University Press, 2001. 45–60. Print.

Read, John, N. Haslan, L. Sayce, and E. Davies. "Prejudice and Schizophrenia: A Review of the 'Mental Illness is an Illness Like Any Other' Approach." *Acta Psychiatrica Scandinavica* 114.5 (2006): 303–318. Print.

Robins, Robert S., and Jerrold M. Post. "Choosing a Healthy President." *Political Psychology* 16.4 (1995): 841–860. *JSTOR.* PDF file. 1 March 2010.

Rosenthal, Andrew. "Dukakis Releases Medical Details to Stop Rumors on Mental Health." *New York Times* 4 August 1988: n. pag. Web. 5 March 2010.

Rountree, Clark. "The (Almost) Blameless Genre of Classical Greek Epideictic." *Rhetorica* 19.3 (2001): 293–305. Print.

Sackett, Russell. "Positive v. Negative in Tom Eagleton Story." *The Capital Times* 27 July 1972. Web. *Newspaper Archive.* PDF file. 5 March 2010.

Salinger, Pierre. "Four Blows That Crippled McGovern's Campaign." *Life* 29 Dec. 1972: 71+. Print.

Samuels, Ellen. "My Body, My Closet: Invisible Disability and the Limits of Coming Out Discourse." *GLQ* 9.1–2 (2003): 233–255. Print.

Schnittker, Jason. "An Uncertain Revolution: Why the Rise of a Genetic Model of Mental Illness Has Not Increased Tolerance." *Social Science & Medicine* 67 (2008): 1370–1381. *ScienceDirect.* PDF file. 10 March 2010.

"Sen. Eagleton's Past." *Chicago Tribune* 27 July 1972: n. pag. Web. *ProQuest Historical Newspapers.* PDF file. 1 April 2010.

Siebers, Tobin. "Disability in Theory: From Social Constructionism to the New Realism of the Body." *American Literary History* 13.4 (2001): 737–754. Print.

Sluiter, Ineke, and Ralph M. Rosen, eds. *Kakos: Badness and Anti-Value in Classical Antiquity.* Mnemosyne: Supplements. History and Archaeology of Classical Antiquity Vol. 307. Leiden/Boston: Brill, 2008. Print.

Soreff, Stephen M., and Patricia H. Bazemore. "Mental Health and the Run for the White House." *Behavioral Healthcare* 28.9 (2008): 20–21. Web. 17 February 2010.

Sprague, Rosamond Kent. "Dissoi Logoi or Dialexis." *Mind* 77.306 (April 1968): 155–167. *JSTOR.* PDF file. 12 March 2010.

Stassen-Berger, Rachel E., and Baird Helgeson. " 'People Have a Right to Know,' Dayton Says." *Star Tribune* 27 December 2009. Web. 2 April 2010.

Sturdevant, Lori. "Mark Dayton Reveals a Private Struggle: Depression." *Star Tribune* 27 December 2009. Web. 2 April 2010.

White, Theodore. *The Making of the President 1972.* New York: Atheneum, 1973. Print.

Wilkinson, Francis. "The Running Mate Who Wasn't." *New York Times* 30 December 2007: n. pag. Web. 17 February 2010.

Wills, Garry. "McCain's McGovern Moment." *New York Times* 3 September 2008. N. pag. Web. 4 March 2010.

Wood, Amy L., and Otto F. Wahl. "Evaluating the Effectiveness of a Consumer-Provided Mental Health Recovery Education Presentation." *Psychiatric Rehabilitation Journal* 30.1 (Summer 2006): 46–52. *Academic Search Complete.* PDF file. 9 March 2010.

The Genre of the Mood Memoir and the *Ethos* of Psychiatric Disability

Katie Rose Guest Pryal

Recent rhetorical accounts of mental illness tend to suggest that psychiatric disability limits rhetorical participation. This article extends that research by examining how one group of the psychiatrically disabled—those diagnosed with mood disorders—is using a particular narrative genre to engender participation, what I call the mood memoir. *I argue here that mood memoirs can be read as narrative-based responses to the rhetorical exclusion suffered by the psychiatrically disabled. This study employs narrative and genre theory to reveal mood memoirists' tactics for generating ethos in the face of the stigma of mental illness.*

The psychiatrically disabled have long suffered exclusion from public life. Historically, doctors have isolated the psychiatrically disabled in asylums, because "doctors believed that between 70 and 90 percent of insanity cases were curable, but only if patients were treated in specially designed buildings" (Yanni 1); courts have imposed on them forced sterilization, believing that "heredity plays an important part in the transmission of insanity" (Buck 206). Today, the psychiatrically disabled continue to be denied civic participation: they are dismissed as criminals, committed patients, or simply unreliable observers of their world.[1] In short, the psychiatrically disabled are not trusted to exercise reason or judgment; as a consequence, civic exclusion often yields rhetorical exclusion as well. For

---

[1]Courts continue to determine whether a mentally ill person is competent—reasonable enough—to stand trial, to testify, to be held responsible for past actions, or to be committed to an institution against her will. These determinations are largely based on the person's ability to participate in judicial proceedings, taking into account factors such as "whether the defendant: (1) is oriented as to time and place; (2) is able to perceive, recall, and relate; (3) has an understanding of the process of the trial and the roles of judge, jury, prosecutor, and defense attorney" among other factors (Hermann 236). Given how few of the sane laypersons involved in the U.S. legal system could provide an adequate "understanding of the process of [a] trial" or other proceeding, this standard of mental competency seems quite high.

example, the "character" portion of the professional licensing process for lawyers scrutinizes psychiatric treatment in order to determine whether a person is competent to advocate in court, despite showings that such scrutiny is both misdirected and harmful (Langford 1220). Doctors often bar psychiatric patients from speaking to the decision-making portion of their treatment, even when treatment has successfully rendered patients able-minded.

Psychiatric disabilities—even mild ones, or ones that respond well to treatment—mark a person as unreasonable or incapable of rational thought, producing an unreliable *ethos* for the mentally ill. Only recently have rhetorical scholars begun to address these rhetorical limitations. Cynthia Lewiecki-Wilson's work investigates "the problem of granting rhetoricity to the mentally disabled: that is, rhetoric's received tradition of emphasis on the individual rhetor who produces speech/writing, which in turn confirms the existence of a fixed, core self, imagined to be located in the mind" (157). As Catherine Prendergast aptly observes, "To be disabled mentally is to be disabled rhetorically" (202). These accounts suggest that psychiatric disability limits rhetorical participation. This essay extends these considerations of rhetorical disability by examining how one particular group of the psychiatrically disabled—those diagnosed with mood disorders[2]—seeks to overcome rhetorical exclusion by way of a narrative genre, what I call the *mood memoir*. I argue here that mood memoirs can be read as narrative-based responses to rhetorical exclusion suffered by the psychiatrically disabled. What is more, mood memoirs can be classified as a genre because their shared exigencies have given rise to shared rhetorical conventions, including an apologia, a moment of awakening, criticism of doctors, and certain techniques of *auxesis* (or rhetorical amplification). Using these shared conventions, mood memoirs have been successful in creating a reliable *ethos* for the mentally ill, at least in certain spheres. For example, the mood memoir genre has gained extraordinary popular appeal in recent years. Many mood memoirs have become best-sellers, such as journalist Elizabeth Wurtzel's *Prozac Nation*; novelist William Styron's *Darkness Visible*; and psychiatry professor Kay Redfield Jamison's *An Unquiet Mind*. In 2002, MTV aired a documentary special titled "True Life: I'm Bipolar," featuring Lizzie Simon, a mood memoirist whose narrative I examine herein.

This study employs narrative and genre theory to reveal mood memoirists' tactics for generating *ethos* in the face of the stigma of mental illness. Mood memoirs employ narrative-based modes of persuasion, entering the rhetorical realm as "stories competing with other stories constituted by good reasons" (Fisher 2). The mood memoir competes with a variety of other stories: those produced by

---

[2]Mood disorders, a category of psychiatric disability recognized in the *Diagnostic and Statistical Manual* (DSM) of the American Psychiatric Association, include depression, bipolar disorder, and other illnesses whose primary symptoms are a disturbance of mood. For a discussion on the Mood Disorder group in the fifth edition of the DSM, see the Report of the Mood Disorders Work Group of the American Psychiatric Association (Fawcett).

doctors, by law-makers, and by popular media in its portrayals of the mentally ill. Furthermore, the mood memoir's "good reasons" can be studied as generic conventions, as I demonstrate below. In the end, then, this article suggests that narratology and genre theory are useful tools for examining the rhetoric of psychiatric disability, and neurorhetorics more broadly, for narrative is an important form through which lay populations come to understand disability. Narratives, that is, serve a rhetorical purpose (not just a literary or aesthetic one), in that they constitute mental illness.

To this end, this article will focus on mood memoirs published by major publishing houses.[3] I further limit my study to the writings by those diagnosed with what the *Diagnostic and Statistical Manual* of the American Psychiatric Association (DSM) calls "mood disorders":[4] depression and bipolar disorder (manic-depression) and all of the sub-categories of these diagnoses.[5] After a review of relevant current rhetorical scholarship, I examine the interaction between narrative and the rhetorical exclusion of the mentally ill, tracing the exigencies in which mood memoirs arise. Next, I highlight some of the generic conventions of the mood memoir as demonstrated by the texts I have examined. Lastly, I point out some implications of the rhetorical space claimed by mood memoirists and discuss the possible limitations of the *ethos* they create.

Recent Rhetorical Examinations of Disability and Psychiatry

Scholars have recently turned to the rhetorical construction of psychiatric disability. Rhetoric of science scholars, for example, have examined this construction through the documents generated by medical professionals—most notably, by the DSM. Stuart Kirk and Herb Kutchins, in their book-length study of the manual, highlight the DSM's role in constructing mental illness through its categorization of symptoms, or its nosology. These scholars reveal just how much psychiatric diagnoses rely on discursive constructions of illness. The rhetoric of the DSM is central to the mood memoir genre, as nearly all memoirists I consider here draw from the book's authoritative force (Jamison *Unquiet Mind* 87; Hornbacher

---

[3]For a closer study of mental illness narratives published in nontraditional, and perhaps more democratic, fora, see Jones.

[4]This limitation fits this study because most psychiatric memoirs on the market today are written by authors with depression and manic-depression, and because authors often use the strong correlation between mood disorders and creativity to justify the writing of a memoir in the first place. There are, however, some well-known memoirs of mental illness whose authors fall outside of the mood disorder category, such as Susanna Kaysen's *Girl Interrupted* (1994) and Elyn R. Saks's *The Center Cannot Hold: My Journey Through Madness* (2007). Kaysen's diagnosis was borderline personality disorder, and Saks's was schizophrenia.

[5]The DSM currently lists five diagnoses under the category of Mood Disorder: Major Depressive Disorder, Dysthymic Disorder, Bipolar I Disorder, Bipolar II Disorder, and Cyclothymic Disorder.

Madness 7; Styron *Darkness Visible* 53–54). Like Kirk and Kutchins, Berkenkotter illustrates how diagnostic criteria, alongside patient narratives, create a means of interpreting symptoms of psychiatric disability—and of billing for treatment. Disability scholars take as their focus the rhetorical disability that often attends mental or psychiatric disability. Following the work of Foucault, Lewiecki-Wilson highlights the rhetorical double bind of mental disability.[6] To achieve rhetorical participation, one's disability must be made invisible, which has the effect of "diluting the transformative potential of their participation in the public forum" (159).

Recently, other scholars have noted the rhetorical power of narrative—and of memoir in particular—in constructing disability. In "Conflicting Paradigms: The Rhetorics of Disability Memoir," disability theorist G. Thomas Couser notes, "Most literary scholars would agree that autobiography has served historically as a sort of threshold genre for other marginalized groups"; however, "disability may disqualify people from living the sorts of lives that have traditionally been considered worthy of autobiography" (78). The disabled lives that often yield published memoirs hardly represent the norm: rather, they are often narratives of the exceptional cases who have overcome disability, of the "Supercrip, who is by definition atypical" (80). The rhetoric of these memoirs "tends to remove the stigma of disability from the author, leaving it in place for other individuals with the condition in question" (80). The mood memoir should be read as a memoir with a different purpose than the typical disability memoir, as it does not track the narrative conventions described by Couser. For example, the mood memoirist rarely seeks to remove stigma through a Supercrip narrative; rather, the memoirist tends to embrace her illness as not just a disability, but also a gift, building an *ethos* based on links between mood disorders and creativity drawn by recent scientific research, as well as upon the historical conception of the mad genius that dates back to Plato.[7]

Mood memoirs can be read as narrative-based responses to rhetorical exclusion suffered by the psychiatrically and mentally disabled. Mood memoirists seize the rhetorical authority provided by the DSM (as described by Kirk and Kutchins)

---

[6]Foucault points out the connection between language and exclusion of the insane: "From the depths of the Middle Ages, a man was mad if his speech could not be said to form part of the common discourse of men. His words were considered null and void, without truth or significance, worthless as evidence, inadmissible in the authentication of acts or contracts" (*Archaeology* 217). The "mad" were thus considered to be utterly without reason; the words of the insane lacked *ethos*, because they could not be relied on to reflect reality.

[7]Plato, in the *Phaedrus*, outlines the divisions of "divine madness" through the voice of Socrates: "The divine madness was subdivided into four kinds, prophetic, initiatory, poetic, erotic, having four gods presiding over them" (140). In the *Ion*, he argues that poets compose from within a trance of madness: "For the poet is a light and winged and holy thing, and there is no invention in him until he has been inspired and is out of his senses, and the mind is no longer in him" (11).

and combine it with the authority provided by autobiography as a genre (as described by Couser). They use narrative to describe their illnesses, often echoing the descriptions of illness that doctors provide in their patient files, seizing the authority inherent in the act of describing for themselves—implying that patients, not just doctors, are capable of describing symptoms and their attendant suffering. Much of the rhetorical authority generated by the mood memoir relies on narrative and the generic conventions of memoir itself.

The narrative nature of the texts mood memoirists produce are ideally suited for their rhetorical purposes of removing taboo from mood disorders, of talking back to the medical profession, and of generating a stronger *ethos* for the psychiatrically disabled. Rhetoricians have often pointed out that narrative has a persuasive purpose. As Donald Phillip Verene writes, "What holds a philosophy together is its narrative aspect. The narrative it expresses is its life blood that animates its arguments and gives them interest" (143). Indeed, for Cicero and Quintilian, the *narratio* of a speech was required before the rhetorician could put forward any arguments. The *narratio* created a shared tale that bound the audience to the speaker, a framework within which the audience might be persuaded. Because it functions to bind audiences and speakers, narrative is central to the "mad movement," the activist movement that seeks to gain acceptance for the mentally ill. MindFreedom International (MFI), a large organization in the movement, features on its web site many stories of "psychiatric survivors," such as that of Leah Harris, "a second generation psychiatric survivor, [who] discovered MindFreedom in 2000 when she was 25 years old. Her first act in the mad movement was to tell her story of oppression and resistance, and to help edit stories for MindFreedom's Oral History Project" (MindFreedom.org). That storytelling was Harris's first "act" in the "movement" implies that narrating one's experience with psychiatry is a form of political activism. The narratives written by mood memoirists should also be read as such: they are political counter-narratives to the dominant psychiatric narratives about mental illness.

Despite efforts of organization such as MFI, the mentally ill continue to be excluded from rhetorical participation on a variety of fronts. These exclusions create the exigencies which give rise to the mood memoir as a genre, a narrative form that, in Fisher's terms, functions as "a dialectical synthesis" of persuasion and literature (2). Narratives, and memoirs in particular, can intercede rhetorically on behalf of people such as the mentally ill who are traditionally excluded from rhetorical participation. Narrative authority—authority grounded in *stories supported by reasons*—provides a way to speak back to experts. In this way, as Fisher observes, narrative can have a democratizing force, albeit one that is still dependent on hierarchies of knowledge and power (9). The mood memoir should be seen as a counter-narrative to a dominant narrative of mental illness put forward by the psychiatric profession. As a group of counter-narratives grows, the narratives tend to adhere to a particular generic form. Judy Z. Segal observes, "People do not fashion their narratives out of just the events of their lives; narratives are structured using available narrative knowledge" (6). Narratives are deeply embedded in the context in which they arise—a context full of other memoirs—such that dominant narratives can easily take hold, often to the exclusion of other, alternative narratives.[8] In fact, as Brett Smith and Andrew Sparkes point out, our cultural context "has a ready stock of narratives from which [storytellers] draw" (18). Authors of counter-narratives constantly contribute to this stock, modifying the rhetorical choices that are available to future authors. These "stock narratives" might best be examined using a theory of genres, to examine the conventions shared across different sets of narratives (say, mood memoirs or cancer narratives) and by narratives within a certain set.

The Mood Memoir as Genre

Using narrative to convey their experiential knowledge, mood memoirists overcome rhetorical exclusion by creating a narrative space in which their voices can be heard, generating a rhetorical capital grounded in narrative logic rather than scientific logic. Mood memoirs, when viewed as a set of narratives, share common features which allow them to be classified as a genre. As Northrop Frye has observed, "The purpose of criticism by genres is not so much to classify as to clarify . . . traditions and affinities, thereby bringing out a large number of literary relationships that would not be noticed as long as there were no context established for them" (247–248). The mood memoir, then, shares traditions and affinities with other narrative genres that have been developed in response to different sorts of rhetorical exclusions. As Tzvetan Todorov observes, genres come "[q]uite simply from other genres. A new genre is always the transformation of an earlier one, or of several: by inversion, by displacement, by combination" (15). In keeping with Carolyn R. Miller's assertion that genre should be viewed "as social action," as "a complex of formal and substantive features that create a particular effect in a given situation" (153), the mood memoir fits within a tradition of other narrative genres designed to generate rhetorical authority for their authors.

For Miller, a genre consists of conventions that have rhetorical consequences; it is "a point of connection between intention and effect, an aspect of social action" (153). A genre arises in response to the exigencies of a situation; motivated by those exigencies, a genre can in turn transform the situation. At other points in U.S. history, memoir genres have arisen to generate rhetorical authority for their authors. For example, during the era of American slavery, the slave-narrative genre served the cause of abolition by giving authority the previously enslaved to speak

---

[8]Ellen Barton points out that, when personal storytelling meets disability activism, the "preferred counter-narrative is not always the narrative told by many individuals with disabilities" (96).

about their experiences. The slave narrative—a genre itself built upon the early-American captivity-narrative genre, as literary scholar John Sekora notes (486)—shares many qualities with the mood memoir. For example, both slave narratives and mood memoirs usually begin with an apologia, in which the author strikes a tone of humility, apologizing for sharing a story of suffering and stating a desire to help others. Mood memoirs and slave narratives also share the convention of awakening. As Sekora explains, antebellum slave narratives tend to describe the autobiographer's escape from slavery and a coming to political consciousness (493). This coming-to-consciousness is reflected in the mood memoir at the moments when the authors experience an awakening to the fact that they are, in fact, mentally ill, and realize that they most likely need psychiatric treatment. The awakening to the reality of illness and treatment represents a move toward metaphorical freedom (from illness) and enlightenment (to a mind that is reasonable, no longer insane).

The awakening moment of the mood memoir also reflects another historical, transatlantic memoir genre: the spiritual autobiography. This genre was often employed by those who experienced religious persecution to gain authority for themselves and for their faith. As noted by literary scholars such as Robert Bell, the conventions of the spiritual autobiography often include a narration of "youthful transgressions" followed by a narrative of religious conversion (112). A distinct parallel exists in the mood memoir, when the memoirist recounts the horrors of life before the moment of awakening to the illness, followed by the attendant diagnosis and effective treatment that occurs post-awakening.

That the mood memoir shares features with these earlier memoir genres suggests that the mood memoir has grown out of familiar autobiographical genres of the past. Recognizing that a "genre, whether literary or not, is nothing other than the codification of discursive properties" (Todorov 15), and recognizing further that a genre is never a "closed set," but rather "an open class, with new members evolving, old ones decaying" (Miller 153), I posit the mood memoir as a new autobiographical genre. The mood memoir genre is motivated by the rhetorical exclusion of the psychiatrically disabled based on their supposed lack of reason. This exigency creates the prime intention—and effect—of the mood memoir: the creation of an authoritative and reliable *ethos* for the author and for others that suffer from mood disorders for the purpose of overcoming this rhetorical exclusion. Mood memoirs move toward this goal through a number of shared conventions. Each mood memoirist considered here

1. discusses in an *apologia* the motivations for writing the memoir, to justify the project and defend it from detractors;
2. experiences a *moment of awakening* to the existence of the illness;
3. criticizes "bad" doctors;
4. lays claim to other sufferers of mood disorders in order to normalize the illness and amplify (via *auxesis*) the memoirist's authority.

The next section will consider how these four conventions are shared across mood memoirs, and what rhetorical purposes they serve.

Convention 1: Apologia

Most mood memoirs contain an apologia, usually in the introduction or afterword, justifying or defending their project. In delivering apologias, "rhetors respond to threats against their 'moral nature, motives, or reputation' by adopting defensive postures of absolution, vindication, explanation, or justification" (Downey 42). Memoirists of all stripes have defended their work from the charge of self-absorption: critics suggest that writing an entire book about oneself is bold, crass, arrogant, or selfish. For example, the apologia was a "stock device" in slave narratives (DeCosta-Willis 8). In the preface to *Incidents in the Life of a Slave Girl*, Harriet Jacobs declares, "I have not written my experiences in order to attract attention to myself" (5). Instead, she hopes "to arouse the women of the North to a realizing sense of the condition of two millions of women at the South, still in bondage" (5). Jacobs thus claims humility by stating that her purpose is to help others, rather than to aggrandize herself. Olaudah Equiano makes a similar apologia at the opening of his slave narrative, *The Interesting Narrative of the Life of Olaudah Equiano*: "Permit me, with the greatest deference and respect, to lay at your feet the following genuine narrative; the chief design of which is to excite in your august assemblies a sense of compassion for the miseries which the Slave-Trade has entailed on my unfortunate countrymen" (192). The apologias of these slave narratives defend the authorship of former slaves and justify their autobiographical projects by casting the authors' purposes as selfless.

Similarly, many mood memoirists draw on the apologia to depict their work as selfless, driven by a desire to break down taboos, which will help other sufferers of mental illness. Kay Redfield Jamison presents an exemplary apologia in the introduction to her book, drawing authority from her scientific background. It would seem that Jamison, Professor of Psychiatry at Johns Hopkins University, would not need to include an apologia. Her ethos is already well-established through her other publications, which include the respected medical text *Manic-Depressive Illness* in 1990 (as co-author with Frederick K. Goodwin), and the groundbreaking *Touched with Fire: Manic-Depressive Illness and the Artistic Temperament* in 1993. Her mood memoir, *An Unquiet Mind: A Memoir of Moods and Madness* (1995), is unique among mood memoirs in that it is written by a psychiatric professional. Even so, Jamison's apologia first notes how, in her clinical work, she has tried to break down taboos surrounding bipolar disorder: "Through writing and teaching I have hoped to persuade my colleagues of the paradoxical core of this quicksilver illness that can both kill and create; and, along with many others, have tried to change public attitudes about psychiatric illnesses in general and manic-depressive illness in particular" (7). In this passage, she makes reference to her earlier work on creativity and manic-depression.[9] Jamison also identifies two audiences she has sought to persuade. Her earlier "writing and teaching" were written primarily for her "colleagues"; in saying so, she implies that the audience of her mood memoir is the lay public. That the mood memoir is written for a broader audience is confirmed when Jamison states that her mood memoir is meant to demystify bipolar disorder and "change public attitudes." By identifying a specific purpose—demystification—and specific audiences, Jamison comes across as a thoughtful and generous person who seeks to help others with both her scientific research and memoir writing. Jamison here cultivates the virtue of *eunoia*, or goodwill, with her audience.

In her apologia Jamison expresses fear of exposing her own experiences with a taboo illness, and works to justify this exposure: "I have had many concerns about writing a book that so explicitly describes my own attacks of mania, depression, and psychosis, as well as my problems acknowledging the need for ongoing medication. Clinicians have been, for obvious reasons of licensing and hospital privileges, reluctant to make their psychiatric problems known to others" (7). She acknowledges the complexity of practicing psychiatry and suffering from psychiatric illness: "It has been difficult at times to weave together the scientific discipline of my intellectual field with the more compelling realities of my own emotional experiences" (7). Here, Jamison invokes a dichotomy between scientific knowledge (*logos*) and emotional knowledge (*pathos*): she points to her "scientific disciplines" as one source of knowledge, and her "emotional experiences" as another. Jamison purports to unite these two types of knowledge in her memoir: "And yet it has been from this binding of raw emotion to the more distanced eye of clinical science that I feel I have obtained the freedom to live the kind of life I want, and the human experiences necessary to try and make a difference in public awareness and clinical practice" (7). Her authority as a mood memoirist stems from her status as both a doctor and a sufferer; by invoking these roles, she frames herself as ideally suited for demystifying mental illness, "mak[ing] a difference in public awareness and clinical practice" (7), the primary project of her memoir as established in her apologia.

Novelist William Styron's apologia justifies his use of memoir—a deviation from his typical fictional mode—as a selfless project meant to help others. He states that he wishes to break down taboos with his memoir, *Darkness Visible: A Memoir of Madness*.[10] He writes that "prevention of many suicides will continue to be hindered until there is a general awareness of the nature of this pain," that

---

[9]The DSM uses the term *bipolar disorder* to refer to illnesses that cause both depression and mania. Jamison uses the term *manic-depressive illness*. The term *manic-depression* also refers to the same illness. I use these terms interchangeably. Referring to a person as a "manic-depressive" (or simply a "depressive"), however, seems unnecessarily limiting of a person's identity, so I avoid such nomenclature.

[10]Styron is the author of many award-winning novels, including *The Confessions of Nat Turner* (1967) and *Sophie's Choice* (1979).

is, the pain of depression (33). Styron tells readers that he first spoke out publicly about his own depression in a *New York Times* Op-Ed piece regarding the suicide of Italian author Primo Levi. After Levi died, many "worldly writers and scholars, seemed mystified by Levi's suicide, mystified and disappointed" (32). For these scholars, Levi's suicide "demonstrated a frailty, a crumbling of character" (32–33). Angered by what he perceived to be a grave misunderstanding of the nature of depression and suicide, Styron wrote his column: "The argument I put forth was fairly straightforward: the pain of severe depression is quite unimaginable to those who have not suffered it, and it kills in many instances because its anguish can no longer be borne" (33). The large number of positive responses to this piece surprised Styron, he writes: "I had apparently underestimated the number of people for whom the subject had been taboo, a matter of secrecy and shame" (33–34). He claims to have written *Darkness Visible* in part because of this response. His apologia thus justifies the otherwise self-centered memoir mode by arguing that the goals of the book are generous and selfless. In this way, Styron, too, cultivates *eunoia*.

A nearly identical story lies behind *Prozac Nation* by Elizabeth Wurtzel, as she describes in her apologia (located in the afterword of the paperback edition). In her apologia, Wurtzel explains that she decided to write the book, in part, because of the responses to an article she published on depression in *Mademoiselle* a few years before (354). After the article was published, she "received a ton of letters . . . which led [her] to believe this was a worthy topic, one that could have widespread resonance in book form" (354). Thus, after breaking the taboo with smaller articles, Wurtzel and Styron write that they were convinced—by reader response—of the worthiness of book-length projects. The responses of other sufferers ostensibly created the exigency that drove their mood memoirs. By including these statements in their memoirs, Wurtzel and Styron present themselves as selfless, countering any presumptions a reader may have held that the authors of memoirs are self-absorbed.

In their apologias both Wurtzel and Styron use the metaphor of the closet in order to better describe their goals of breaking down taboos. Wurtzel writes: "In effect, if *Prozac Nation* has any particular purpose, it would be to come out and say that clinical depression is a real problem, that it ruins lives, that it ends lives, that it very nearly ended *my* life; that it afflicts many, many people, many very bright and worthy and thoughtful and caring people" (356). Here, Wurtzel invokes the metaphor of the closet when she employs the term *come out*. Styron writes, regarding his *Times* piece, "The overwhelming reaction made me feel that inadvertently I had helped unlock a closet from which many souls were eager to come out and proclaim that they, too, had experienced the feelings I had described" (34). Fear of exposure has forced other sufferers to hide their illnesses, and the authors of mood memoirs claim a desire to help others to stop hiding. In fact, for Styron, the motivation of helping others is the only one he describes: "I thought that . . . it would be useful to try to chronicle some of my own experiences with the illness and in the process perhaps establish a frame of reference out of which one or more valuable conclusions might be drawn" (34). Similarly, mood memoirist Lizzie Simon writes, "I want to produce a new image for bipolar people. I want to present new voices of bipolar people" (41). This justification based on selflessness is pervasive throughout the mood memoir apologias.

Another common justification that arises in the apologias comes in the form of a comparison. Memoirists often compare their illnesses to more socially acceptable diseases in order to convince readers of the seriousness of mood disorders, usually diabetes (Hornbacher 71; Styron 9) and cancer (Jamison 102; Styron 9, 33; Wurtzel 21). By proving that their disease is serious and deadly, the authors further justify writing memoirs about their experiences with the disease.

Other motivations besides justification manifest in the apologias, such as a compulsion to write and a desire for therapeutic release, for catharsis. For example, Wurtzel describes a compulsion to write her memoir, as though the story of her madness was itself driving her mad: "I had tried very hard to get away from thinking or feeling depression in all of my professional endeavors, but it just kept creeping up, over and over again, like a palimpsest, a text hiding beneath whatever else I was working on that refused to remain submerged" (355). In order to gain release from her thoughts about depression, Wurtzel explains, "I gave in to the obsessive hold that my experiences with depression seemed to have on me, and decided to just write a whole book" (355). Wurtzel thus suggests that the creative drive overcame her own wants and desires. Wurtzel uses the justification of catharsis to counter the charge of self-absorption by suggesting that she did not have the option to *not* write the memoir.

Jamison, too, describes an emotional drive to write her memoir, but hers is less of a compulsion or desire for therapeutic release, and more of a desire for honesty. Despite fears of negative ramifications on her career, Jamison decides to tell her story, suggesting that "the consequences...are bound to be better than continuing to be silent" (7). She is "tired of hiding, tired of misspent and knotted energies, tired of the hypocrisy, and tired of acting as though I have something to hide. One is what one is, and the dishonesty of hiding behind a degree, or a title, or any manner and collection of words, is still exactly that: dishonest" (7). Jamison wishes to step out of the closet described by Wurtzel and Styron.

For Marya Hornbacher,[11] author of *Madness: A Bipolar Life* (2008), writing her mood memoir allows her to reclaim power over her life, a therapeutic goal:

> How do we know who we are or what we can become? We tell ourselves
> stories. The stories we tell are what we know of ourselves. We are a creation,

---

[11]Hornbacher is also the author of *Wasted: A Memoir of Anorexia and Bulimia* (1998) and a novel, *The Center of Winter* (2005).

a product of our own minds, a pastiche of memory, dream, fear, desire. My
memory looks like a child's collage, or a ransom note, incomplete and full of
holes. All I have is today, this moment, to work with. I am writing my story
as I go. (278)

Thus, Wurtzel, Jamison, and Hornbacher all describe writing as more of a need
than a desire. They are fervid, engaged, and committed to their writing. In these
passages, they construct their *ethos* not from reason but rather from emotional
drive and creativity. They mark their invention as passionate, rather than reasoned. They claim to be driven by their creative or emotional urges. Further, they
tap into the correlation between creative genius and mood disorders drawn by
psychiatric researchers in order to gain authority as writers, invoking a dichotomy
between rational and scientific sources of knowledge (*logos*) and emotional and
creative sources of knowledge (*pathos*); it is the second that they claim to draw
upon in their narratives.

Convention 2: Awakening

A second convention of the mood memoir is the moment of awakening to the
reality of the memoirist's illness. Mood memoirs often include an early denial
of the illness and then an awakening, followed by the confession of the illness
to others and the seeking of treatment, and treatment failures followed by a
final success. These conventions share strong similarities to the conventions
of older memoir genres, such as the spiritual autobiography (or conversion
narrative), when the author comes to God; the slave narrative, when the former
slave comes to political consciousness; and the breast cancer narrative, when
the "survivor" discovers a lump (Segal 4). The similarities are so striking that,
as I suggest above, the mood memoir can be thought of as a recent addition to
a much older family.

During the awakening moment of the mood memoir, the memoirist points to a
specific moment in which she first realizes she is mentally ill. The mood memoirist's awakening is unique among similar genres in that the narrator is recounting
the awareness of a mental illness; an illness which, in a paradoxical fashion, does
not impede the author's self-awareness. This apparent paradox renders the awakening moment as particularly rhetorically powerful: the precision of the descriptions of awakening establishes the author's reliability as narrator, despite the
author's mood disorder. By describing the awakening with ostensible clarity, precision, honesty, and self awareness, the memoirist tacitly invites the audience to
trust her narration of events. In short, although the mood memoir is a story of
mental illness, the awakening implies that the illness does not impede the narrator's ability to tell the story.

The awakening is both the turning point of the author's life and the impetus
for the narrative. Mood memoirs thus often commence with the moment of

awakening, and then "swoop"[12] back in time to tell of the period of sickness and denial which occurs before. For example, Styron's memoir opens with this awakening: "In Paris on a chilly evening late in October of 1985 I first became fully aware that the struggle with the disorder in my mind—a struggle which had engaged me for several months—might have a fatal outcome" (3). By describing his awakening to the "disorder in [his] mind," Styron implicitly declares that he is capable of describing this disorder accurately—after all, if he can become aware of the illness in the first place, he must be capable of recounting the illness for readers. After recounting how he came to realize the risk of suicide while driving down the Champs-Elysées, he writes, "Only days before I had concluded that I was suffering from a serious depressive illness, and was floundering helplessly in my efforts to deal with it" (5). Styron had come to Paris to accept a prestigious award for his writing. The moment of awakening is thus tied to what should have been a positive event in his life. Styron's story suggests that that the positive nature of the award highlighted his suffering, making it easier to perceive. The "self-loathing," he writes, caused him to feel "persuaded that [he] could not be worthy of the prize, that [he] was in fact not worthy of any of the recognition that had come [his] way in the past few years" (19). After the award ceremony, he rushes home on the Concorde to seek psychiatric treatment. At this point, the narrative swoops back in time to tell of how Styron grew sick and finally reached the nadir of his disease in Paris.

Lizzie Simon describes her awakening at the beginning of her memoir, *Detour: My Bipolar Trip in 4-D*: "What started the day after I found out I was accepted to college was an episode so horrific that it would become impossible for me to deny that I had a mental illness for the rest of my life. Though I had always known that something was wrong with me, what started that day was evidence, concrete evidence" (3). Simon uses the term "concrete evidence" to emphasize both the concrete nature of the illness (an illness that many still consider to be merely a character flaw) and the concrete nature of her recollection—a recollection that readers can rely upon to be accurate. Like Styron, Simon's awareness comes on the heels of a positive life event, her acceptance to Columbia University, which throws her illness into high contrast. And like Styron's memoir, the next chapter swoops back in time, to Simon's birth in 1976, and wends its way forward to the moment of awakening.

Like Styron and Simon, Jamison begins her book with an awakening tied with an otherwise positive life event: "Within a month of signing my appointment papers to become an assistant professor of psychiatry at the University of California, Los Angeles, I was well on my way to madness; it was 1974, and I was twenty-eight years

---

[12]The "swoop" is a common narrative structure in short fiction. It entails commencing a story with the turning-point of the narrative, then going back in time to tell the relevant events leading up to the turning-point, and lastly moving forward in time to the climax of the story and the end.

old" (4). Jamison then reaches back in time to discuss life events that eventually led to her mental illness, beginning with her childhood. Sixty pages later, the narrative returns to the awakening: "I did not wake up one day to find myself mad. Life should be so simple. Rather, I gradually became aware that my life and mind were going at an ever faster and faster clip until finally, over the course of my first summer on the faculty, they both had spun wildly and absolutely out of control . . . . In the beginning, everything seemed perfectly normal" (68). She then pinpoints a precise moment when she realized she was ill: "Although I had been building up to it for weeks, and certainly knew something was seriously wrong, there was a definite point when I knew I was insane" (82–83). Like Simon's "concrete evidence," Jamison describes a "definite point," emphasizing not only the reality of the illness but the reliability of her recollection of it.

Hornbacher first awakens to her bipolar disorder during a visit with an insightful psychiatrist. After the psychiatrist delivers the diagnosis, Hornbacher writes, "My chest floods with a mixture of horror and relief. The relief comes first: something in me sits up and says, *It's true*. He's right, he has to be right. This is it. All the years I've felt tossed and spit up by the force of chaos, all that time I've felt as if I am spinning away from the real world, the known world, off in my own aimless orbit—all of it, over . . . . I have a word. *Bipolar*" (66–67). Hornbacher emphasizes the "rightness" of the diagnosis; she depicts the diagnosis as a lodestar, a precise point that will guide her out of her "aimless orbit." For the reader of her memoir, the medical term arises out of the chaos of her disjunctive narrative, providing a clear framework to interpret the chaotic behavior Hornbacher describes. Hornbacher's description of awakening is a moment of sublime clarity, one that the reader can share as well.

Whether they are describing a moment of awakening or criticizing doctors (as I describe below), mood memoirists rely on experiential knowledge—transmitted via narrative—to provide an alternative source of knowledge about mood disorders for their readers. This alternative source of knowledge, often rhetorically opposed to more traditional medical or scientific knowledge, creates a space in which memoirists can speak with authority, despite their diagnoses and (in most cases) lack of medical training.

Convention 3: Criticizing Doctors

A third convention of the mood memoir is criticism of doctors and other psychiatric care providers. Nearly all mood memoirists write about interactions with doctors who ignore patient stories in favor of other forms of knowledge, such as observations and diagnostic criteria. When they criticize doctors, mood memoirists pass judgment on the failures of the medical system, relying upon their experiential knowledge to talk back to traditional medical knowledge. This "talking back" gives rhetorical power to the memoirists; they deliberately oppose their experiential, narrative-based knowledge as patients to the empirical, scientific knowledge of the medical profession. Doctors, like lawyers and other experts, can be blinded to the different types of knowledge possessed by the lay. For example, in the debates over "regressive" autism, the experiential, narrative-based knowledge of parents slam against the nearly united scientific front of the medical community.[13] For the mood memoirist, casting the psychiatric establishment as monolithic gives their position—their opposition—more power. When memoirists talk back to the medical profession to regain authority, contrasting the narrative of experiences with the lack of insight demonstrated by health care providers, the source-of-knowledge dichotomy employed by mood memoirists—*pathos* versus *logos*—is particularly striking. Mood memoirists claim to rely upon emotional, experiential knowledge, transmitted through an artistic, nonlinear narrative, in order to speak against the *logos* of the medical profession.

Marya Hornbacher presents a striking example of the medical profession's failure: she suffered from a life-threatening eating disorder throughout her childhood and, as a consequence, was often misdiagnosed with depression and prescribed anti-depressants. Psychiatrists agree that anti-depressants are generally thought to cause mania in persons with bipolar disorder (Cipriani and Geddes para. 6). Hornbacher writes: "No one even thinks *bipolar*—not me, not any of the many doctors, therapists, psychiatrists, and counselors I've seen over the years—because no one knows enough. Later, this will seem almost incredible, given what a glaring case of the disorder I actually have and have had nearly all my life" (7). Here, she suggests that because her case was "glaring," a typical presentation of bipolar disorder, the medical professionals who treated her failed in their treatment. Implied in her language is that, even Hornbacher, a non-specialist, can see that her symptoms constitute a diagnosis of bipolar. This criticism, then, points to weaknesses in the presumption of expertise and strengthens the position of the non-expert. Later in her life, after she attempts suicide by cutting open a vein in her forearm, the psychiatrist releases her from the hospital because "Hospital policy is to impose the least level of restriction possible" (5). Hornbacher states that she is "very convincing" when she promises she will not injure herself again. The suicide attempt occurred in 1994. It is not until 1997 that she met the doctor who diagnoses her with bipolar disorder. This narrative forms the basis of Hornbacher's argument that current treatment of bipolar disorder is seriously flawed.

Jamison, too, roundly criticizes medical professionals involved in her case, recounting an interaction with a doctor who asked her whether she planned to have children (*Unquiet Mind* 190). She responded in the affirmative, and, "At that point, in an icy and imperious voice that I can hear to this day, he [the doctor] stated—as though it were God's truth—which he no doubt felt it was—'You shouldn't have children. You have manic-depressive illness'" (191). She reflects

---

[13]For more on the autism debates and the diverse forms of proof employed by the parties, see "A Broken Trust: Lessons from the Vaccine-Autism Wars" by biologist Liza Gross.

further, "Brutality takes many forms, and what he had done was not only brutal but unprofessional and uninformed. It did the kind of lasting damage that only something that cuts so quick and deep to the heart can do" (191). From her position of authority as a clinical scientist, Jamison criticizes the doctor's knowledge and professionalism. From her position as a mood memoirist, she criticizes his behavior using emotional appeals grounded in metaphorical terms: "brutality," "cuts," and "heart." These emotional appeals help pinpoint the weaknesses of medical treatment of bipolar disorder, lending more authority to Jamison as a mood memoirist—and, in a unique fashion—as an insider (in that she is a psychologist) stepping outside (as a patient) to critique standard practices of psychiatry.

Styron describes many bad doctors in his narrative. One, Dr. Gold, failed to "say much of value" (*Darkness Visible* 53). Instead, "His platitudes were not Christian but, almost as ineffective, dicta drawn straight from the pages of the *Diagnostic and Statistical Manual of the American Psychiatric Association* (much of which, as I mentioned earlier, I'd already read), and the solace he offered me was an antidepressant medication called Ludiomil" (53–54). In this passage, Styron tries to generate *ethos* by relying on a medical text—the DSM. He strikes at the authority of the doctor by suggesting that the doctor did nothing more than what he, Styron, could have done—read the DSM and applied the diagnostic criteria. This argument attempts to demystify (and thereby disempower) the psychiatric profession.

Styron later describes a moment when the same doctor gave Styron the anti-depressant: "Dr. Gold said with a straight face, the pill at optimum dosage could have the side effect of impotence" (*Darkness Visible* 60). Here, Styron presents a doctor performing ostensibly correct medical treatment: warning a patient of side effects of a new drug. But Styron argues that the correct treatment is actually flawed: "Until that moment...I had not thought him totally lack in perspicacity; now I was not at all sure.... I wondered if he seriously thought this juiceless and ravaged semi-invalid with the shuffle and the ancient wheeze woke up each morning from his Halcion sleep eager for carnal fun" (*Darkness Visible* 60). Styron's narrative, placed in opposition to the correct treatment provided by Dr. Gold, seeks to reveal weaknesses inherent in what the profession accepts as correct treatment. Dr. Gold, Styron argues, was a very poor observer and listener; implied in this claim is that doctors should improve these skills. In this scene, Styron, the patient, was shocked that the doctor could so misperceive his physical symptoms. Styron illustrates the disjuncture that can grow between the narratives patients tell doctors, the doctors' interpretations of these narratives, and the treatments doctors prescribe based on these interpretations.

Later, Styron criticizes Dr. Gold once more for advising against hospitalization (67–68). Styron writes, "Many psychiatrists, who simply do not seem to be able to comprehend the nature and depth of the anguish their patients are undergoing, maintain their stubborn allegiance to pharmaceuticals in the belief that eventually the pills will kick in, the patient will respond, and the somber surroundings of the hospital will be avoided" (68). Styron relies on his experiential knowledge—that a hospital did indeed cure his depression in the end—to speak against the theoretical knowledge of doctors. His narrative provides the evidence needed to support these claims.

Convention 4: Laying Claim

The fourth convention that I examine here I call *laying claim*: every mood memoirist I have studied lays claim to other sufferers of mental illness as a means to create rhetorical authority. By "laying claim" I mean, on one level, simple name- or statistics-dropping. Memoirists seek to normalize their illnesses; until quite recently—and many would argue, still—patients were advised to hide their diagnoses. Memoirists can break this taboo by pointing out the large number of people who are also diagnosed with mental illness (statistics-dropping) or by invoking famous persons with mental illness (name-dropping). The rhetorical function of name-dropping, then, looks something like this: Here is another person who suffers from mental illness like me; this person is respected, so I should be respected too. A greater number of sufferers suggest that although mental illness can be horrible, it is not unusual.

Laying claim has deep roots in the history of rhetoric. In the *Rhetoric*, Aristotle makes suggestions for amplifying the reputations of those praised in epideictic rhetoric. He writes, "[I]f you cannot find enough to say of a man himself, you may pit him against others" (I *Rhetoric* 9, Trans. Roberts, 33). The rhetor should compare the object of praise "with famous men; that will strengthen [the rhetor's] case; it is a noble thing to surpass men who are themselves great" (I *Rhetoric* 9, Trans. Roberts, 33). Aristotle calls this practice of comparison "amplification,"[14] or *auxesis* in Greek. Since mental illness is often taboo, it is not surprising that laying claim—comparing oneself to a famous person who has a similar illness— would take central importance in establishing authority.

Laying claim becomes complicated, however, when a claimed famous person is long dead and never actually diagnosed with a mood disorder. In order to lay claim to such a person, the memoirist must perform a retroactive diagnosis of the famous person's supposed mental illness. Often, the memoirist is not the first to speculate about the person's illness. If the famous person committed suicide, or kept a diary recording mental suffering, or produced fiction or art portraying it, scholars and biographers have often already pointed to mental illness as a root cause. Such retroactive diagnoses place the memoirist in a tradition of others with mood disorders; laying claim to great artists, such as Van Gogh, Plath, and Woolf, populates this tradition with creative geniuses. In summary, laying claim has two

---

[14]The Roberts translation, whose text I quote, uses the phrase "heightening the effect"; Kennedy's translation uses the term "amplification."

purposes: it can either normalize the author—a great step in establishing authority—or, through retroactive diagnosis of creative geniuses, push the author's *ethos* into the realm of genius, which seems like an even stronger rhetorical position, particularly in the context of creative nonfiction.

In Lizzie Simon's memoir, unique among the ones I studied, laying claim is the central project. Simon was diagnosed with bipolar disorder shortly before starting college. She was compliant as a patient, accepting medication and psychiatric care and living a life of full civic participation. After graduating from Columbia, she worked as an assistant producer in a successful theater. She conceived of a project to travel the country to interview other young people with bipolar disorder. These people are not famous—except through Simon's project—and like Simon participate fully in society. In her afterword, she explains the reasons for her project: "Everybody I interviewed for this book is diagnosed with bipolar affective disorder, between the ages of sixteen and thirty-five, on medicine, and highly functional in society. We do not share the same illness, for we each experience it differently. But we do share the same diagnosis" (210). Here, Simon's words assert that bipolar disorder is prevalent, given the large cross-section of people she interviewed. She also asserts that the disorder is not monolithic, because "each [person] experience[s] it differently." But they do share a diagnosis of bipolar, and furthermore, she writes, they "share the same nagging inner voice that wonders: how much of me is me, and how much of me is the illness?" (210). Emphasizing the experiences that they *do* share helps Simon lay claim to the experiences of the people she interviewed. They do, as Simon points out, share the same experience of alienation: "This is our interior, private response to the exterior, public noise of stigma" (210). By emphasizing the commonness of the disorder and the commonality of the experiences of the people diagnosed with it, Simon attempts to break down this stigma. Simon's project was eventually produced as a television special on MTV, titled "True Life: I'm Bipolar," one more step in the process of normalization.

Another form laying claim takes in a memoir is that of simple numbers, or statistics-dropping. For example, Marya Hornbacher has a list of "Bipolar Facts" at the end of her book, which starts with this statistic: "American adults who have bipolar disorder: 5.8 million (2.8% of the U.S. population)" (*Madness* 281). The list of statistics emphasizes the prevalence of bipolar disorder and suicide, normalizing them. In the prologue to her book, Jamison writes: "The disease that has, on several occasions, nearly killed me does kill tens of thousands of people every year: most are young, most die unnecessarily, and many are among the most imaginative and gifted that we as a society have" (*Unquiet Mind* 5). Here, Jamison attempts to remove the stigma from her suicide attempts and diagnosis by highlighting that both bipolar disorder and suicide are far more common than most people suspect.

Jamison's use of the adjectives "imaginative and gifted" point to the second function of laying claim: locating oneself within a tradition of creative geniuses. The creative genius trope is a powerful source of authority, one that most mood memoirists employ. In order to invoke the creative genius trope, mood memoirists perform retroactive diagnoses of famous artists and writers. Early in her memoir, Wurtzel writes, "I'm starting to wonder if I might not be one of those people like Anne Sexton and Sylvia Plath who are just better off dead, who may live in that bare, minimal sort of way for a certain number of years, may even marry, have kids, create an artistic legacy of sorts, may even be beautiful and enchanting at moments" (*Prozac Nation* 8). Although her tone is one of depressed self-deprecation ("better off dead"), she associates herself with two towering geniuses of poetry. Later in the narrative she lays claim again, noting how, when she first suffered from mental illness, "The idea that a girl in private school in Manhattan could have problems worth this kind of trouble seemed impossible to me. The concept of white, middle-class, educated despair just never occurred to me.... I didn't know about Joni Mitchell or Djuna Barnes or Virginia Woolf or Frida Kahlo yet. I didn't know there was a proud legacy of women who'd turned over-whelming depression into prodigious art" (50). With these words, Wurtzel places herself firmly within that proud legacy of artistic genius.

Of the memoirs considered here, Styron's most prominently features retroactive diagnosis. First, he mentions Albert Camus and his disappointment at never meet-ing him. Styron, a writer located in the upper echelons of fame and prestige, actu-ally had an appointment to meet Camus at a dinner party in Paris with a mutual friend named Romain Gary. But the meeting never occurred, because, Styron writes, "before I arrived in France came the appalling news: Camus had been in an automobile crash, and was dead at the cruelly young age of forty-six.... I pon-dered his death endlessly... there was an element of recklessness in the accident that bore overtones of the near-suicidal" (*Darkness Visible* 22). Styron bolsters his retroactive diagnosis of Camus's suicidal tendencies by pointing to suicide in Camus's writings and by citing information passed on to him from their mutual friend: "Camus, Romain told me, occasionally hinted at his own deep despon-dency and had spoken of suicide" (23). Styron acknowledges, however, that the car in which Camus died was driven by someone else.

Styron also lays claim to his friend, author Romain Gary, "twice winner of the Prix Goncourt" (28), a highly prestigious French literary award, who committed suicide after lunch one day. Gary "went home to his apartment on the rue du Bac and put a bullet in his brain" (28). In addition to Camus and Gary, Styron lays claim to Abbie Hoffman (29), Randall Jarrell (30–31), Primo Levi (32), John Dryden (44), Sir Walter Scott (44), the Brontes (44), Emily Dickenson (45), Baudelaire (46), and many others. But Styron is hardly the only memoirist to use retroactive diagnosis to borrow the authority of creative geniuses. Hornbacher lays claim to Byron, and attributes her "career and passions" to her illness (279); Jamison lays claim to composer Hugo Wolf (39), Virginia Woolf (68), and Edna St. Vincent Millay (73), among others. By locating themselves in a legacy of creative geniuses, mood memoirists—who are, after all, practicing a creative art in writing their memoirs in the first place—grant themselves greater authority as writers. They can show that one can achieve great things even with, or because of, a mood disorder.

The convention of laying claim has been further complicated by the medical research of Jamison and other authors, who have shown correlations between mood disorders and the "artistic temperament" (Jamison, *Touched with Fire*). This research, combined with the fine creative work produced by many mood memoirists, has created a popular misconception that all mood disorders carry creative potential. For example, as Jonah Lehrer recently observed in the *New York Times Magazine*, depression may have an "upside." Lehrer ponders the prevalence of depression in the human population, and suggests that "depression has a secret purpose and our medical interventions are making a bad situation even worse" (para. 7). He points to Aristotle and Keats (Lehrer para. 16), and to researchers who seek to show how rumination caused by depression "might lead to improved outcomes, especially when it comes to solving life's most difficult dilemmas" (Lehrer para. 17). "For Darwin," according to Lehrer, "depression was a clarifying force" (para. 4). This perceived connection between creative genius (such as Darwin's) and mood disorder lends a rhetorical authority grounded in creativity to *all* who would write a mood memoir. However, this connection is highly controversial, as the furor over Lehrer's article demonstrates.[15] Although research shows that there might be a correlation between creative genius and mood disorders, the causation—that is, the notion that mood disorders *cause* creativity—is not at all clear.

Thus, rhetorical authority based on the creativity trope (which itself is often built on weak historical and scientific evidence), seems to serve many mood memoirists' stated purposes of removing the taboo from mental illness and helping other sufferers to be more open about their diagnosis. But this type of authority might have implications outside of the literary/artistic sphere in which the memoirists are claiming authority. After all, there are many people diagnosed with mood disorders who might not wish to relinquish an *ethos* grounded in reason or science. In many ways, the retroactive diagnosis of artists is a weakness of the mood memoir genre. Not only is the practice often medically unsound[16]— Styron's presumption of Camus's "suicide" is particularly striking in this regard—but it does little to break down the traditional rhetorical limitations of the mentally ill outside of literary or creative spheres. It is possible that a diagnosis

---

[15]Lehrer's article prompted a spate of articles in response to his position; many authors were supportive of Lehrer, and many others were deeply troubled by his conclusions. See, for example, Eric Jaffe of *Psychology Today*, Ronald Pies at *PsychCentral.com*, and James Gordon at *The Huffington Post*.

[16]The problematic retroactive diagnosis of mental illness echoes Foucault's work, especially on the retroactive labeling of the ancient Greeks as "homosexual." Foucault historicizes the term—and concept—of homosexuality, pointing to "Westphal's famous article of 1870 on 'contrary sexual sensations,'" as the "date of birth" of the "medical category of homosexuality" (*History of Sexuality* 43). As homosexuality itself did not exist before 1870, it is logically unsound to discuss same-sex behavior as "homosexual" if it occurred in a historical location before Westphal's rhetorical establishment of homosexuality.

of a mood disorder might not affect one's career as a writer or artist. After all, much scientific work has been produced drawing a correlation between artists and mood disorders. However, if you work in a field that does not prize artistic genius—if you are a judge or scientific researcher, for example—this authority grounded solely in the mystique of a creative genius does you little good.

Conclusion

Rhetoric scholars have effectively illustrated how the mentally ill have been excluded from rhetorical participation. I suggest here that rhetoric scholars should start examining the ways that the mentally ill attempt to move past this exclusion, in particular, how they employ narrative to speak against dominant perceptions of mental illness. Narrative genres provide quasi-democratic fora in which oppressed groups—such as those with neurological difference—can gain rhetorical authority. This study of the generic conventions shared across sets of narratives, and within a particular set (such as the mood memoir), makes more apparent the rhetorical challenges faced by oppressed groups.

For example, in this study, I have shown that a mood memoirist gains rhetorical authority through a variety of methods: by declaring a selfless purpose for writing the memoir in an apologia; by describing the moment of awakening to an illness; by speaking back to bad doctors; and by laying claim to others with mood disorders in order to normalize a diagnosis. The declaration of selflessness in the apologia certainly fits within a tradition of apologias shared across various sets of narrative genres; but it also reveals that mood memoirists, as a set, are aware of the stigma attached to psychiatric disability and that they seem to desire to break down that stigma. The description of the moment of awakening reveals a memoirist's desire to establish an *ethos* of a reliable narrator—of a reliable observer of experiences. By speaking back to doctors, memoirists cast the medical profession as complicit in their rhetorical exclusion, and hope to counter this exclusion by revealing medicine's weaknesses. Lastly, the convention of laying claim serves to counter the isolation and weaken the stigma that often accompanies diagnoses. In summary, this study shows how a particular approach to neurorhetorics, one that examines narratives and the generic conventions they share, provides a new perspective on the rhetorical exclusion suffered by those with neurological differences.

References

American Psychiatric Association. *Diagnostic and Statistical Manual of Mental Disorders: DSM IV-TR [Electronic Resource].* Arlington, VA: American Psychiatric Publishing, 2000. *Psychiatryonline.com.* Web. 7 March 2010.

Aristotle. Rhetoric. W. Rhys Roberts. Adelaide, Australia: eBooks@Adelaide, The University of Adelaide Library, 2010. Web. 1 July 2010. <http://ebooks.adelaide.edu.au/a/aristotle/>.

———. *On Rhetoric: A Theory of Civic Discourse.* Trans. George Alexander Kennedy. Oxford University Press, 1991. Print.

Barton, Ellen. "Disability Narratives of the Law: Narratives and Counter-Narratives." *Narrative* 15.1 (2007): 95–112. *ProjectMuse*. Web. 1 July 2010.

Bell, Robert. "Metamorphoses of Spiritual Autobiography." *ELH* 44.1 (1977): 108–126. *JStor*. Web. 25 June 2010.

Berkenkotter, Carol. "Genre Systems at Work: DSM-IV and Rhetorical Recontextualization in Psychotherapy Paperwork." *Written Communication* 18.3 (2001): 326–349. *Sage Journals Online*. Web. 7 March 2010.

Buck v. Bell. 274 U.S. 200. Supreme Court of the U.S. (1927). *Justia.com U.S. Supreme Court Center*. Web. 22 April 2010.

Cipriani, Andrea, and John R. Geddes. "Antidepressants for Bipolar Disorder: A Clinical Overview of Efficacy and Safety." *Psychiatric Times* 25.7 (2008). *Psychiatrictimes.com*. Web. 1 July 2010.

Couser, G. Thomas. "Conflicting Paradigms: The Rhetorics of Disability Memoir." In *Embodied Rhetorics: Disability in Language and Culture*. Eds. James C. Wilson and Cynthia Lewiecki-Wilson. Carbondale, IL: SIU Press, 2001. 78–91. Print.

DeCosta-Willis, Miriam. "Self and Society in the Afro-Cuban Slave Narrative." *Latin American Literary Review* 16.32 (1988): 6–15. *JStor*. Web. 1 July 2010.

Downey, Sharon D. "The Evolution of the Rhetorical Genre of Apologia." *Western Journal of Communication* 57.1 (1993): 42–64. *ERIC*. Web. 28 April 2010.

Equiano, Olaudah. "The Interesting Narrative of the Life of Olaudah Equiano, or Gustavas Vassa, the African." *My Soul Has Grown Deep: Classics of Early African-American Literature*. Ed. John Edgar Wideman. Philadelphia: Running Press, 2001. 192–368. Print.

Fawcett, Jan. Report of the Mood Disorders Work Group. *American Psychiatric Association*. 2008. *Psych.org*. Web. 21 April 2010.

Fisher, Walter R. "Narration as a Human Communication Paradigm: The Case of Public Moral Argument." *Communication Monographs* 51.1 (1984): 1–22. *Informaworld*. Web. 1 July 2010.

Foucault, Michel. *The Archaeology of Knowledge*. New York: Pantheon Books, 1972. Print.

———. *The History of Sexuality: An Introduction, Volume 1*. Trans. Robert Hurley. New York: Vintage, 1990. Print.

Frye, Northrop. *Anatomy of Criticism: Four Essays*. Princeton University Press, 1971. Print.

Goodwin, Frederick K., and Kay R. Jamison. *Manic-Depressive Illness: Bipolar Disorders and Recurrent Depression*. 2nd Ed.. Oxford University Press, 2007. Print.

Gordon, James. "Depression's Upside: A One-Sided View." *The Huffington Post*. 11 March 2010. *HuffingtonPost.com*. Web. 1 July 2010.

Gross, Liza. "A Broken Trust: Lessons from the Vaccine–Autism Wars." *PLOS Biology* 7.5 (2009): e1000114 1–14. *Plosbiology.org*. Web. 22 April 2010.

Hermann, Donald H. J. *Mental Health and Disability Law in a Nutshell*. St. Paul, MN: West Publishing Company, 1997. Print.

Hornbacher, Marya. *Madness: A Bipolar Life*. New York: Mariner Books, 2009. Print.

———. *The Center of Winter*. New York: HarperCollins, 2005. Print.

———. *Wasted: A Memoir of Anorexia and Bulimia*. New York: HarperCollins, 1998. Print.

Jacobs, Harriet. *Incidents in the Life of a Slave Girl*. Eds. Nellie Y. McKay and Frances Smith Foster. New York: W.W. Norton & Co., 2001. Print.

Jaffe, Eric. "Depression's Upside, Down." *PsychologyToday.com*. 4 March 2010. Web. 1 July 2010.

Jamison, Kay Redfield. *An Unquiet Mind: A Memoir of Moods and Madness*. New York: Alfred A. Knopf, 1995. Print.

———. *Touched with Fire: Manic-Depressive Illness and the Artistic Temperament*. New York: Free Press, 1996. Print.

Jones, Raya A. "Identity Commitments in Personal Stories of Mental Illness on the Internet." *Narrative Inquiry* 15.2 (2005): 293–322. *Ingentaconnect*. Web. 1 July 2010.

Kaysen, Susanna. *Girl, Interrupted*. New York: Vintage, 1994. Print.

Kirk, Stuart A. and Herb Kutchins. *The Selling of DSM: The Rhetoric of Science in Psychiatry*. Hawthorne, NY: Aldine de Gruyter/Transaction, 1992. Print.

Langford, Carol M. *"Barbarians at the Bar: Regulation of the Legal Profession Through the Admissions Process." Hofstra Law Review* 36.4 (2008): 1193–1225. *HeinOnline*. Web. 1 July 2010.

Lehrer, Jonah. "Depression's Upside." *New York Times* 28 February 2010: MM, 38+. *NYTimes.com*. Web. 1 July 2010.

Lewiecki-Wilson, Cynthia. "Rethinking Rhetoric through Mental Disabilities." *Rhetoric Review* 22.2 (2003): 156–167. *JSTOR*. Web. 7 March 2010.

Miller, Carolyn R. "Genre as Social Action." *Quarterly Journal of Speech* 70.2 (1984): 151–167. *EBSCOhost*. Web. 7 March 2010.

MindFreedom International. "Mad Pride—Welcome !" *MindFreedom.org*. 2010. Web. 1 July 2010. <http://www.mindfreedom.org/campaign/madpride>.

MTV. "True Life: I'm Bipolar." Air Date: 25 July 2002. Television.

Pies, Ronald. "The Myth of Depression's Upside." *PsychCentral.com*. 1 March 2010. Web. 1 July 2010.

Plato. *Ion*. Trans. Benjamin Jowett. Adelaide, Australia: eBooks@Adelaide, The University of Adelaide Library, 2010. Web. 1 July 2010. <http://ebooks.adelaide.edu.au/p/plato/>.

Plato. *Phaedrus*. Trans. Benjamin Jowett. Adelaide, Australia: eBooks@Adelaide, The University of Adelaide Library, 2010. Web. 1 July 2010. <http://ebooks.adelaide.edu.au/p/plato/>.

Prendergast, Catherine. "On the Rhetorics of Mental Disability." In *Towards a Rhetoric of Everyday Life: New Directions in Research on Writing, Text, and Discourse*. Madison: University of Wisconsin Press, 2003. 189–206. Print.

Saks, Elyn R. *The Center Cannot Hold: My Journey Through Madness*. New York: Hyperion, 2007. Print.

Segal, Judy Z. "Breast Cancer Narratives as Public Rhetoric: Genre Itself and the Maintenance of Ignorance." *Linguistics and the Human Sciences* 3.1 (2008): 3–24. *Equinoxjournals.com*. Web. 1 July 2010.

Sekora, John. "Black Message/White Envelope: Genre, Authenticity, and Authority in the Antebellum Slave Narrative." *Callaloo* 32 (1987): 482–515. *JStor*. Web. 1 July 2010.

Simon, Lizzie. *Detour: My Bipolar Road Trip in 4-D*. New York: Washington Square Press, 2003. Print.

Smith, Brett and Andrew Sparkes. "Narrative and Its Potential Contribution to Disability Studies." *Disability & Society* 23.1 (2008): 17–28. *Informaworld*. Web. 1 July 2010.

Styron, William. *The Confessions of Nat Turner*. New York: Random House, 1967. Print.

———. *Darkness Visible: A Memoir of Madness*. New York: Vintage, 1992. Print.

———. *Sophie's Choice*. New York: Random House, 1979. Print.

Todorov, Tzvetan. *Genres in Discourse*. Trans. Catherine Porter. New York: Cambridge University Press, 1990. Print.

Verene, Donald Phillip. "Philosophy, Argument, and Narration." *Philosophy & Rhetoric* 22.2 (1989): 141–144. Print.

Wurtzel, Elizabeth. *Prozac Nation*. New York: Riverhead Books, 1995. Print.

Yanni, Carla. *The Architecture of Madness: Insane Asylums in the United States*. Minneapolis: University of Minnesota Press, 2007. Print.

Index